How to Successfully Encourage Sustainable Development Policy

This book builds on the insights of Günther Bachmann, former Secretary General for the German Sustainability Council, who spent almost 20 years advising the German government on sustainability policy. The book discusses his experiences in negotiating stakeholder statements at the highest national levels.

Bachmann takes the reader behind the scenes of German sustainable policy and practice, whilst also comparing Germany with other national approaches. He tells the story of political events from his insider perspective, unfolding the narrative of sustainable development goals and how activists in their respective countries could and should relate to it. Furthermore, he suggests new lines of vision through the tangle of conference fatigue and buzzwords. The book argues that environmentalists often display entrenched attitudes that too often downplay success. The rhetoric of crisis and doom, if overstretched and reduced to alarm, paralyses action and innovation. Bachmann, who on the contrary argues positively and concretely, shows unusual but significant signs of hope and confidence in action and how these can be made effective in the politics of sustainable development.

This book will be of global relevance to sustainability professionals and policy makers and will advise them on how to successfully move ahead with sustainability policies.

Günther Bachmann served as Secretary General for the German Sustainability Council from 2001 to 2020, advising the German federal government and Chancellor Angela Merkel on sustainability policies. As Secretary General, his duties included the initiation, guidance and implementation of all content and process-related work of the council. Günther also conducted and facilitated three international peer reviews on the German sustainability strategy, chaired by Björn Stigon (2008 and 2013) and Helen Clark (2018), and established innovative formats with international resonance such as the corporate Sustainability Code or the SDGclub.Berlin. Günther is Innovators Fellow at Conservation International and advises the Columbian think tank CEPEI. He sits on the executive board of the German Sustainability Awards and has chaired the award's jury for companies and municipalities since 2008. He is a member of the scientific advisory board of the Thünen Institute for Rural Areas, Forestry and Fisheries. In addition, he is a member of the board of trustees of Nabu, the German Nature Conservation Association (Bird Life) and is partner of a consultancy.

Routledge Studies in Environmental Policy

Environmental Policy in India
Edited by Natalia Ciecierska-Holmes, Kirsten Jörgensen, Lana Ollier and D. Raghunandan

Mainstreaming Solar Energy in Small, Tropical Islands
Cultural and Policy Implications
Kiron C. Neale

EU Environmental Governance
Current and Future Challenges
Edited by Amandine Orsini and Elena Kavvatha

The European Union and Global Environmental Protection
Transforming Influence into Action
Edited by Mar Campins Eritja

Environmental Policy and Air Pollution in China
Governance and Strategy
Yuan Xu

Climate Change Law and Policy in the Middle East and North Africa Region
Edited by Damilola S. Olawuyi

European Foreign Policy in a Decarbonising world
Challenges and Opportunities
Sebastian Oberthür, Dennis Tänzler, and Emily Wright, with Gauri Khandekar

How to Successfully Encourage Sustainable Development Policy
Lessons from Germany
Günther Bachmann

For more information about this series, please visit: www.routledge.com/Routledge-Studies-in-Environmental-Policy/book-series/RSEP

How to Successfully Encourage Sustainable Development Policy

Lessons from Germany

Günther Bachmann

Routledge
Taylor & Francis Group
LONDON AND NEW YORK

earthscan
from Routledge

First published 2022
by Routledge
605 Third Avenue, New York, NY 10158

and by Routledge
4 Park Square, Milton Park, Abingdon, Oxon, OX14 4RN

Routledge is an imprint of the Taylor & Francis Group, an informa business

Library of Congress Cataloging-in-Publication Data
A catalog record for this book has been requested

ISBN: 978-1-032-14974-5 (hbk)
ISBN: 978-1-032-14952-3 (pbk)
ISBN: 978-1-003-24202-4 (ebk)

DOI: 10.4324/9781003242024

Typeset in Times New Roman
by Apex CoVantage, LLC

Contents

Figures

Figures designed by © Bert Odenthal, Berlin

Acknowledgements

The content of this book is based on my practical experiences at the front lines of the transformation from deeply fossil- and growth-bound politics to a sustainable society, as policy advisor and network organiser.

In my previous role as Secretary General of the Council for Sustainable Development, I was fortunate to meet leaders, protagonists, and activists and work with them on our common cause. My continuing role with the German Sustainability Award adds assessing corporate solutions.

I owe a great debt of gratitude to all members of the council, a total of some 40 outstanding personalities in German politics. Above all, however, the council chairpersons have urged me to make a difference in German politics together with them or in the name of the council, and to keep thinking anew about ways and means. These were, in the order of their tenures as chairperson, Matthias Platzeck, Volker Hauff, Klaus Töpfer, Hans-Peter Repnick, Marlehn Thieme and Werner Schnappauf. In the first ten years of the council's work, the leadership of Volker Hauff and Klaus Töpfer shaped my work and enriched my understanding. I am grateful that we remain in close contact to this day and manage one political task or another.

I would also like to thank Stefan Bauernfeind and his predecessors in the German chancellery and the chiefs of staff of the chancellery, State Secretaries Hans Martin Bury and Frank-Walter Steinmeyer and Ministers Thomas de Maizière, Ronald Pofalla, Peter Altmaier, Helge Braun for their ambition and engagement. For their continuous support in parliament, I would like to thank MPs Rüdiger Kruse, Andreas Jung, Bettina Hoffmann, Michael Thews and Bernd Westphal, and the parliamentary pioneers Ingrid Arndt-Brauer, Ulla Burchard, Winfried Herrmann, Matthias Miersch, Michael Kauch, Astrid Klug, Lukas Köhler, Günter Krings, Nina Scheer, Jürgen Trittin and Valerie Wilms.

Of course, it was only through the help and support of my team in the council's office that I was able to make good use of budgets and windows of opportunity. They have driven German sustainability policy forward with tireless enthusiasm and professional skill. For this, I would like to express my sincere thanks to all of them.

I also want to thank Stefan Schulze-Hausmann and the team of the German Sustainability Awards. Since 2007, our joint project provides an amazing flow of insights, leverage and good vibes that are so much needed in the community.

Profound international expertise has helped me to better understand risks and weaknesses as well as chances and opportunities. I am very grateful to Adolfo Ayuso-Audry, Gábor Bartus, Arnau Queralt Bassa, Paula Caballero, Helen Clark, Rodrigo Diaz, Felix Dodds, Chrysoula Exarchou, Karl Falkenberg, Elisabeth Freytag-Rigler, Pekka Haavisto, Javier Jurasky, Annika Lindblom, Patti Londoño, Herbert Lust, Louis Meuleman, Namhla Mniki-Mangaliso, Tumainiel Mangi, Mohammed Valli Moosa, Martha Talamondjila Naanda, John Patrick Ngoyi, Ingeborg Niestroy, Alessandra Nilo, Charles Nouhan, Derek Osborn, Joost Oorthuizen, László Pintér, Philipp Schönrock, Nikolaus Schultze, Mike Simson, Cletus I. Springer, Achim Steiner, Björn Stigson, Jan-Gustav Strandenaes, Darren Swanson, Marcel Szábo, Wolfram Tertschnigg, Farooq Ullah, Michiel de Vries and Irena Zubcevic, to name only a few. I also thank those friends and colleagues who repeatedly invited me to give keynotes at the UN Office for Sustainable Development's Transition Forum, other international meetings and across Europe.

Not to forget the first 20 years of my professional activity at the German Federal Environmental Agency, which was dedicated to the protection of soil as a natural resource. I would like to sincerely thank all my former colleagues, superiors and friends for the good time, which was an excellent preparation for my further commitment to sustainability. I thank Munich publisher Oekom for allowing me to reprint parts of my 2020 book *Die Stunde der Politik*.

Abbreviations

ASviS	Italian Alliance for Sustainable Development
BHAG	Big Hairy Audacious Goal
CBDR	Common But Differentiated Responsibilities
CEPEI	Centro de Pensamiento Estratégico Internacional
CO2	carbon dioxide
COP	Conference of the Parties
CCS	carbon capture and storage
CCUS	carbon capture, utilisation and storage
CSR	corporate social responsibility
CsrD	European Corporate Sustainability Reporting Directive
DGNB	Deutsche Gesellschaft für Nachhaltiges Bauen e.V. (Green Building Standard)
DNK	Deutscher Nachhaltigkeitskodex (German Sustainability Code)
EEAC	European Environmental and Sustainable Development Advisory Councils
EFRAG	European Financial Reporting Advisory Group
EGD	European Green Deal
EMAS	Eco-Management and Audit Scheme
ESD	Education for Sustainable Development
ESDN	European Sustainable Development Network
EU	ETS European Union Emission Trading System
GAVI	vaccine alliance; created as global alliance for vaccines and immunisation
GDP	gross domestic product
GHG	Greenhouse Gas
GINI	index named after the statistician Corrado Gini that measures the income inequality within a social group
GSA	German Sustainability Award
GSDS	German Sustainable Development Strategy
HLPF	High-Level Political Forum on Sustainable Development (of the United Nations) under the auspices of UN ECOSOC
IASS	Institute for Advanced Sustainability Studies
IFRS	International Financial Reporting Standards

INTOSAI	International Organization of Supreme Audit Institutions
IPCC	Intergovernmental Panel on Climate Change
IPBES	Intergovernmental Science-Policy Platform on Biodiversity and Ecosystem Services
ISSB	International Sustainability Standards Board, established by the IFRS Foundation
KfW	Kreditanstalt für Wiederaufbau
MDGs	Millennium Development Goals
MSI	multistakeholder initiative
NGO	non-governmental organisation
NCS	natural climate solutions
OECD	Organisation for Economic Cooperation and Development
RENN	Regionale Netzstelle Nachhaltigkeitsstrategie (Regional Hubs Sustainability Strategy)
RNE	Rat für Nachhaltige Entwicklung (German Council for Sustainable Development, Council)
SDGs	Sustainable Development Goals (set by the 2015 United Nations General Assembly)
SDS	sustainable development strategy
TSVCM	Taskforce on Scaling Voluntary Carbon Markets
TWh	Terra watt hours
UNDESA	United Nations Department of Economic and Social Affairs
UNEP	United Nations Environmental Program
UNESCO	United Nations Educational, Scientific and Cultural Organization
UNOSD	United Nations Office for Sustainable Development
VNR	Voluntary National Report (to the UN HLPF)
WBCSD	World Business Council for Sustainable Development

1 Introduction

Never trust a lesson without thinking for yourself

If there is one thing I have learned about lessons and the concept of "lessons learned" it is to stay realistic, and sceptical. The first lesson from any event or book talking about lessons learned should be never trust a lesson, at least, not without further thought. There are two reasons for this: one is of a more general type, the other is more German-specific.

Of course, there is definitely a lot to learn about how to successfully drive sustainability politics. So, the idea of lessons learned and of sharing of experiences has almost become a defining element of any sustainable development conference, maybe more so than in other topics. Politicians and activists never tire of emphasising the need to learn from each other. Entire manuals are giving guidance on how to write lesson learned reports. Conference participants spend hundreds, if not thousands, of hours listening to, presenting or discussing lessons learned. Today, this has become an inflationary must-do that more often than not just occurs as a naive routine. Those endless hours are a part of political correctness, in particular with the social media generation. If every bit that had been offered and published as a lesson had really added value to people's insights and influenced the debate, the world would be better off, one can assume. However, the amount of real time learnings does not correspond to the inflation of rather superficial and "standard" offering of lessons. The actual learning is not keeping pace. That's why I use the concept very carefully.

Second, the breach of civilisation that occurred during the Holocaust forces us Germans to be maximally reluctant about telling others what to do. Suggesting there is anything at all to be learned from the Germans sounds awkward and absurd. For our decent fellow Germans, it is axiomatic that the German *Vergangenheitsbewältigung* (the coping with the past, politically, culturally, legally, as individuals and as society at large), still today, is incomplete. So, restraint became a virtue of German politics after liberation from fascism. It applied in the East as well as in the West. For a long time, German restraint could not be overdone. The moral high ground is not an option for Germany. As the economically strongest member state of the European Union, Germany naturally has a special responsibility to hold Europe together in the face of increasingly divergent interests. This can be done, if at all, with honesty, respect and clarity about one's own interests, not by pushy lecturing. After fascism, against all odds, and with significant help of

DOI: 10.4324/9781003242024-1

other nations, we developed a democracy and succeeded in building a responsible and accountable nation, and now begin the energy transition and a sustainable transformation.

This casts a new and very different light on Germany and certainly is challenging the rigor of the restraint. Susan Neiman, an American who directs the Einstein Forum, a public think tank outside of Berlin, suggests that America and other nations should learn from the German experience of coping with evil. Neiman (2019) sees Germany's culture and politics deeply informed by this history. She is right. Even if this does not completely rid the German reality of racism and anti-Semitism, it seems to keep right-wing populism in check and helps democratic parties resist the bold reflexes of brazen populism. When Helen Clark, who served as prime minister of New Zealand from 1999 to 2008, and was the administrator of the United Nations Development Programme from 2009 to 2017, presented the results of her review of German politics to then-Chancellor Angela Merkel, she said that "what Germany does on sustainability is of global importance, not least because of the size of its economy and population. When major countries move on sustainable development, that is felt around the world" (Clark 2018a). In the review itself the peers allude to the global recognition of Germany's capacity for sustainability, and continue by asking: "If Germany can't pull out all the stops, who could? Those expectations of Germany will most likely increase" (Clark 2018b, p. 11).

People and politicians around the world yearn for democracy, sustainability and human rights (well, maybe not all). All countries should set out on the path to sustainable development and adopt national sustainability strategies. This is what the 2030 Agenda for Sustainable Development called for in 2015, repeating and strengthening what had been discussed as early as 1992 in the Earth Summit UN conference in Rio de Janeiro.

The world has agreed on sustainable development goals and on the target of not warming the Earth more than two degrees Celsius and therefore drastically cutting greenhouse gas emissions by 2050, in the even better case to a maximum of only 1.5 degrees Celsius. Some countries, including Germany, are translating this goal into a political promise to become climate neutral by 2045/2050 and to achieve full decarbonisation where possible. The goals that the global community has signed up to in the run-up to 2030, such as the elimination of hunger and malnutrition, the reduction of food waste, sustainable consumption and industrialisation, and the protection of the oceans, are no less ambitious than the Paris climate target. And they are to be achieved as early as 2030, having been adopted in 2015. Like two flags, 2030 and 2050 mark the commitment of the global community to at least minimally live up to its civilisational self-image. Half of these 15 years are reached in 2021/2022, but not half of the target distance (see Figure 1.1). All countries have a duty to find solutions to keep them in line with the 2030 agenda.

All countries? The differences between the countries of the world and within their societies are serious. As a result, development and equity must be brought together. People in the poorest countries insist on their right to free themselves from the hardships of hunger and poverty. And no one would "deny it to them".

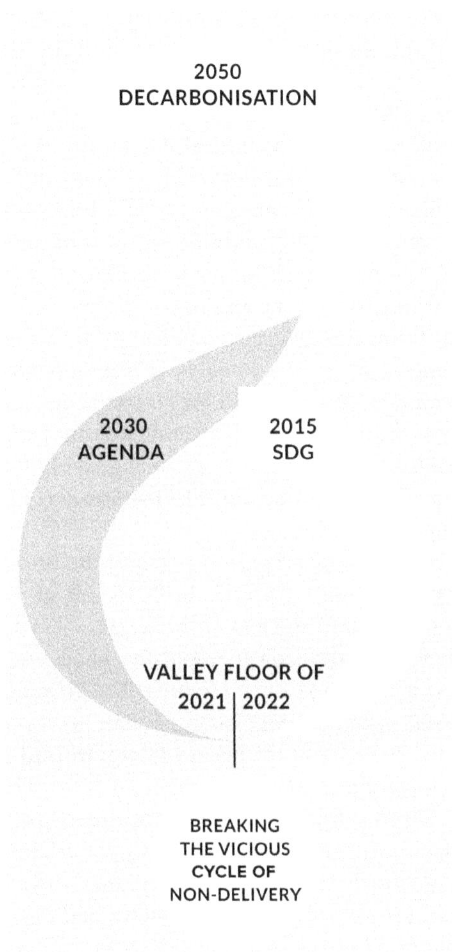

Figure 1.1 The valley floor of 2021/2022

All people everywhere should be able to live a life in dignity while preserving natural resources. This approach tries to make human development sustainable, and for this purpose, it adds "time" (the long run) and intergenerative justice as guiding essentials for politics and policies. This is also imperative for German domestic policy, since ecological and economic domestic policies still tend to neglect the social dimension in their distributional effects.

The major emerging economies have a special position. Without their contribution, a global sustainability turnaround is impossible. High growth rates and their economic dynamism generate far too many climate-damaging emissions and pollutants. For instance, their share of the plastic flood in the oceans is high.

Strangely enough, these countries associate the overcoming of extreme poverty with an increase in the number of super-rich people. It seems the term *emerging economies* can no longer be appropriate. They, too, must undergo fundamental transformation.

The world is demanding that Germany and the other industrialised countries with early (and therefore now old) industrialisation take on a peacemaker role in the turnaround to sustainability. The self-image of German politics is consistent with this role, but still not every German policy finds itself replicated in the world. The reasons for this may be a nitty gritty overengineering, high bureaucracy or unreasonable costs. Another reason may be insufficient communication of the idea, alternatives and solutions on the part of Germany.

Improving policy for sustainable development is an imperative, and lessons learned can be an important tool in doing so if they stimulate further thinking. Ambitious claims and honorary goals are of no use if they are not met. In fact, there is a gap between the ambition and the reality of the Paris Climate Accords. Emissions of greenhouse gases worldwide are rising, not falling. This is the Paris valley, and it is getting bigger and deeper. The pandemic is further widening this gap of vicious non-delivery.

The valley floor is even more serious in terms of the broader United Nations Sustainable Development Goals (SDGs). In 2015, all of the world's nations agreed to the SDGs of the 2030 agenda. This was preceded by extensive discussions and negotiations. The decision to adopt the goals was not taken lightly. All of them are backed up with facts and figures. The 2030 agenda has met with much approval and also generated enthusiasm. Today, however, most of the goals already envisioned for 2030 are in the distant future. Instead of getting closer to them step by step, global trends are taking them off track. This is the reason for very firm warnings, whether from activists on the street, books or celebrities, or even at the highest level from the secretary general of the United Nations. He has declared a decade of action for the 2020s to achieve the 2030 goals and is stepping up the rhetoric. He is certainly right. Commitments and target achievement are clearly mismatched. For all people who believe in the credibility of politics and the rationality of political engagement, this gap is more than a temporary problem. Rather, it is a valley of non-delivery in the pursuit of the global commons.

The simple truth is that the world, while not at ground zero of commitments and goals, is in a zero position in terms of actions to deliver for those goals. The future, for better or worse, is decided on the bottom of this valley. Will we succeed in getting on a path that can redeem even important and complex goals? Or will the ambitious goals become a redevelopment case, getting ever more problematic to clean up? The beginning of the 2020s is crunch time, when the damaging trends are to be broken or the bottom is to be at least passed, or it remains at the bottom for the time being. Even after that, the challenge will remain on the table; sustainability is, after all, a process.

That countries themselves must do more is hardly disputed by those with the highest political responsibility. Only the *how* remains vague and undecided. How to build and strengthen institutions? How to successfully address and engage the private sector? How to "bring people along"? How to build on collective competence and strengthen self-efficacy? How can technical and administrative innovations go hand in hand with a culture of sustainability? How to build trust, and what role free advice can play through bodies with multistakeholders? How dare we hope? These are all serious questions, and open ones at that. They are being asked around the world.

The German approach to sustainability explicitly wants to be open to the world, innovative and capable of globalisation, and to reach people "at home". The German energy transition is the flagship megaproject. With phasing out fossil and nuclear power and phasing in climate neutrality, this transition is a triple North Star. In a way, the ambition of the energy transition may remind us of a core element of a computer server station that is being taken out and replaced, while the server is in full swing. On the scale required, this has never been done before in any old industrialised country, and success is certainly not a given.

On the other hand, preconditions are good, both in terms of cultural preparedness, people's mindsets, technology, and economic feasibility. Interestingly, the idea of sustainability is culturally deeply rooted in Germany and those roots are actively used as a driver for innovation. Germany is good at building institutional capacity, and the political system is strong enough to enable new forms of gatekeeping. No other government puts its own strategy before high-level experts for criticism so openly. After all, no other society that depends so heavily on the industrial economy has been pursuing a sustainability strategy for so long, enabling such extensive networking with the private sector, civil society and academia, and politics at all levels.

Both in climate protection and in sustainability policy, activists, ambitious officials and innovative companies form large and well-connected communities of like-minded people (with all the differences on details). This is good and certainly a prerequisite for overcoming the Paris chasm as well as the SDG chasm. To redeem this prerequisite, these communities need more influence and decision-making power. They also have to muster the courage to go off-script, to connect with deep roots, be visionary, develop a failure culture, understand better what makes a success, and how to create reasons for hope. However, for this to happen, they also need to change "internally". They are still moving mostly in a bubble of their own making. The usual catchphrases there internalise how to address their own clientele. Unpleasant topics are often left out. We see this in Germany. Climate neutrality and a 100-percent supply of renewable energies might even become empty words because there is too little disclosure of the impositions and the limits they entail if everything else is left unchanged. An honest discussion would certainly be better in some places, because it would likely overturn conventional ideas about the implementation of goals, for example. By no means is it the case that good governance is just the product of determination, administrative coherence and alignment. This book presents insights and, from an insider's and practitioner's view, shows where there is a momentum in current policies, what organisational and social learning really means and where there are hidden fields of change. It seeks to provide information always with a critical eye toward unnecessary obstacles and mistakes, but also with pride in what has been achieved, insofar as this is appropriate.

Germany's achievements, critics often say, are nice to look at, but they look very "German": overly precise, too little life, too much engineering. And honestly, they sometimes certainly are. However, what works for Germany is good enough to drive the domestic solution-finding. It comes with a realistic, if not sceptical understanding of "lessons" that those achievements are no textbook solutions. Similarly, some see the phasing out of nuclear energy and coal and achievement of climate neutrality for the entire country by 2045 as overambitious. However, there are great potential benefits should this plan succeed. Critical voices have to take this into account. It could bring great advances in technology and social innovation; it could set a trend. As a concerted action, sustainability strategy could, after all, achieve more than the sum of sectoral measures.

That's the subjunctive. But what about the present of change? In the midst of transformation, it may still seem abstract and distant to some contemporaries, but as a matter of fact, the present demands that we make decisions and take action. In German politics, this is generally understood, and there is also a readiness to act. How do we more efficiently combine the mutual advantages of digitalisation and sustainability? How do we secure and stabilise German supply chains, e.g. in semiconductors and information technology? How does German foreign policy act in the sales and growth markets for countless German companies, not only in the automotive industry? How can and should the European Union develop further? How can Germany strengthen sustainability policy in Africa? How can we make sustainability solutions mandatory without being overly prescriptive? How do we finance the huge investments required for this transition in the face of the debt that piles during the COVID-19 pandemic and the budgetary debt-to-GDP threshold that should not be above the 60% European threshold (the so-called debt brake (*Schuldenbremse*) is a German constitutional provision)? How will financial inflation and the financial debt management play out, and what about the non-financial debts? Will people stay in favour of sustainability if there are no quick fixes?

New and open so-called spaces of possibilities must be created, while the majority of people, in the face of threats, tend to react cautiously, conservatively and delimitatively to themselves. This is a political challenge. It is obligatory for German politics to actively navigate the future transformations, remembering that no future has ever been mastered by those who cling to the past, and no future-bound policy has any value or ethics when it simply forgets about the past. In his international farewell speech as the outgoing president of the Socialist International in 1992, former German Chancellor and leader of the North-South World Commission Willy Brandt, mortally ill, put it like this:

> Nothing comes of its own accord. And only very little is lasting – be aware of your strength and of the fact that every era requires its own answers and that you must feel up to it if you are meant to do good things.
>
> (Brandt 1992)

Examining the conditions for success and failure in Germany and comparing them with those in other countries, motivated by a sense of urgency about the problems at hand, is instructive and enlarges one's own framework for action. This is so not because good practices are easily or directly transferable (they usually are not), but because the exchange enlarges one's own thinking space. In the best of cases, this turns out to yield the seeds of change.

References and recommended reading

Brandt, Willy (1992) *Farewell Message of the Outgoing President of the Socialist International*, www.willy-brandt-biografie.de/wp-content/uploads/2017/08/Grusswort_Willy_Brandt_SI_1992.pdf

Clark, Helen (2018a) *The Report of the International Peer Review of Germany's Sustainable Development Strategy*. Speech Given in Berlin, www.helenclarknz.com/speeches-on-sustainable-development-gender-equality-and-climate-action/the-report-of-the-international-peer-review-of-germanys-sustainable-development-strategy

Clark, Helen (Chair) (2018b) *Change – Opportunity – Urgency: The Benefit of Acting Sustainably*, Rat für Nachhaltige Entwicklung (ed.). The 2018 Peer Review on the German Sustainability Strategy, Berlin, www.nachhaltigkeitsrat.de/wp-content/uploads/2018/05/2018_Peer_Review_of_German_Sustainability_Strategy_BITV.pdf

Federal Government of Germany (2021) *German Sustainable Development Strategy Update 2021 – Summary*, www.bundesregierung.de/resource/blob/992814/1942596/dff76fd102a1e14aafbe53de4f20a462/deutsche-nachhaltigkeitsstrategie-2021-kurzfassung-englisch-bf-download-bpa-data.pdf?download=1

Neiman, Susan (2019) *Learning from the Germans: Race and the Memory of Evil*. New York: Farrar, Straus and Giroux.

2 Germany

The basic facts

2.1 Key economic, ecologic and social data

Germany is a medium-sized country, one of many in the global geography. In the European Union, however, Germany is the by far largest economy and it produces the fourth largest gross domestic product globally (see Figure 2.1).

After the liberation of Germany from fascism in 1945, in the zones occupied by the USA, Great Britain and France, Germans founded the Federal Republic of Germany (FRG), and in the Soviet occupation zone they founded the German Democratic Republic (GDR). Both states saw themselves as the continuity of an all-German state and did not recognise each other. They faced each other on the front lines of the Cold War. With the fall of the Berlin Wall in 1989 came reunification in 1990 and the end of the Cold War. The legal system of the FRG is extended to the former GDR.

Today, 83 million people live in Germany and share an area of 357,582 square kilometres. The dense population is evident from the average of 233 inhabitants per square kilometre. The average life expectancy is around 81 years, and increasing. The demographic shift toward an aging population continues and is only slightly slowed by immigration. Most Germans live in small towns with a population of between 20,000 and 50,000. Jobs in industry and construction account for 29 percent of gross value added. In 2020, the economic value of all exports (export) exceeds the value of imports of 1.03 trillion euros at 1.2 trillion. According to records published by the bank Credit Swiss (2021), Germany holds a share of 4.4 percent of the whole world's wealth.

Germany had over 8,800 nature reserves as of 2017. Their approximately 2.6 million hectares are located on 6.3 percent of the total area. Some rare or extinct animal species could be reintroduced. About 10 percent of the agricultural land is managed according to organic farming standards. Overall, however, the state of the environment in Germany presents a mixed picture. The air in inner cities has become cleaner; waste is treated in an orderly manner and some is recycled. Soil is protected by law. Numerous pollutants are banned or under control. But recycling is incomplete, only seven percent of rivers are in good ecological condition, groundwater is heavily polluted with fertilisers in many places. Biodiversity and landscape quality are deteriorating. On average, 56 hectares of land

DOI: 10.4324/9781003242024-2

→ developing country in terms of sustainability

→ sustainability targets since 2001

→ industrialized

→ 83 million inhabitants

→ densely populated

→ Gini-Index ca 30

→ federal democracy

Figure 2.1 Basics on Germany

are taken out of agricultural use every day to make way for new settlement and traffic areas. For years, this figure fell; recently it is rising again, as is the number of square meters of living space per person. The number of hot days is rising, as is the number of deaths that can be linked to summer heat. Droughts affecting farmland and floods destroying urban settlements and causing death are occurring much more frequently.

The Corruption Perceptions Index (CPI) for Germany is 80 out of 100 points, putting Germany in the top ninth place on the list of 180 countries. The CPI is determined by the non-governmental organisation Transparency International. The German government includes the CPI as an indicator in the German Sustainability Strategy.

The United Nations Development Programme ranks Germany among the countries with very high human development. In the Global Competitiveness Index, Germany ranked third in 2018. Germany is the largest economy in Europe. The economic strength is characterised by an unusually high number of small and medium-sized enterprises in very specialised industries, the high penetration of science and innovation and the still high proportion of industrial jobs for old industrialised economies. Before the pandemic, Germany's national debt was about 61 percent of gross domestic product. Germany is part of the world's largest domestic market with a combined population of around 500 million. The gross domestic product in 2020 is about 3,329 billion euros, and after eight years of continuous growth, it fell by 5 percent, due to the pandemic-induced lockdown. The government has taken on new financial debt as a result of the pandemic. In

2020, the state debt will reach 2332 billion euros, which is equivalent to 70 percent of the gross domestic product. Paying back this enormous sum is certainly a financial burden for the future. But the general view justifies this by the fact that the instruments of the welfare state and social responsibility in the market economy (short-time allowances instead of layoffs, subsidies instead of company closures) have succeeded in keeping the immediate economic damage of the pandemic under control.

Income is distributed relatively equally in Germany. Globally, the level of equality/inequality of income is expressed by an index, as suggested by the statistician and sociologist Corrado Gini. At 29.7 points, the Gini index for Germany in 2019 is below the European average, not to mention the much wider spread of the index when compared on a global scale. Wealth is much more concentrated in Germany, with a Gini index of 0.78. The number of the super-rich grows in line with the growth in share value of major German stock market.

Germany has never thinned out the social safety net as most countries have done following neoliberal doctrine. Financial and other assistance for the needy has been more or less systematically expanded. The social budget in the overall German federal budget includes all social payments that go to pensioners and the unemployed, as well as, for instance, public subsidies for health insurance, reimbursement costs incurring in connection with maternity and continued wage payments in times of an economic crises. This budget grows. The total sum of all parts of the federal social budget is about more than 1,100 billion euros in 2020, with an expected sum of 1,278 billion in 2025. In 1991 the total sum was 395 billion and even at that point in time, the social system was not perceived as dramatically underfinanced. In percentage of the overall budget the social budget totals at 33 percent in 2020, up from 25 percent in 1991. If there ever was an austerity policy (I doubt it), at least now there is no talk of it. The state is pursuing an extremely expansive budget strategy in the pandemic.

The opportunity for a good education is better than ever for young people from all strata of the population. This is the overall figure, but for a significant number of people opportunities are more or less a theoretical thing, and unattainable. Equality of educational opportunity has still not been achieved regardless of income and parental social status. Education for sustainable development is mostly a matter of (numerous) projects and not yet being systematically embedded in the education structure. Gender policy is making progress. Germany is gradually transforming itself into an immigration country.

2.2 Characteristic lines of conversations

> Making development sustainable is to ensure that it meets the needs of the present without compromising the ability of future generations to meet their own needs. . . . Yet in the end, sustainable development is not a fixed state of harmony, but rather a process of change in which the exploitation of resources, the direction of investments, the orientation of technological development, and institutional change are made consistent with future as well as present needs. We do not pretend

> that the process is easy or straightforward. Painful choices have to be made. Thus, in the final analysis, sustainable development must rest on political will.
>
> (Brundtland 1987)

This definition is still the key to the German understanding of sustainable development. The "needs of the present" is the part that is much more controversial than the part about future generations, in Germany. On the one hand, there is talk of targeted impoverishment and social coldness; on the other hand, there is talk of exaggerated welfare and of convenience. International comparative figures such as the Gini index, according to which Germany is better off than generally assumed, hardly play a role in the discussion. As a rule, the opponents of social policy try to exclude sustainability issues and do not address sustainability goals. How social respect and peace can be secured in times of fundamental transformation is a very controversial topic.

Nowadays, the concept of sustainability is cited and discussed at all levels these days. It enjoys a high level of approval and is used for advertising. It is also increasingly appearing in the public print and TV media, where for a long time journalists considered it a virtue not to even mention the word. But that is only the surface. Beneath the surface, the social and economic aspects dominate the debate in one form or another. Main lines of conversation in Germany can be identified:

- First, on growth.
- Second, on functionality.
- Third, on power.
- Fourth, on red versus green.
- Fifth, on angst.

First, the saying "first the chimneys must smoke, then comes environmental protection, if you can afford it" is still very familiar. Even today, it is often said that sustainability is a luxury good and only something for the rich (people, cities, companies, countries) who want to afford it. Some of the climate policy rhetoric, especially that proposing the most ambitious targets, plays along easily suggesting technical solutions plus carbon pricing and promises not to harm growth (green growth as a result of techno fixes plus pricing).

The German discussion is characterised by a technical belief in the category of growth, in two ways. One is that either growth is held to be indispensable and it is given the status of a civilisational achievement without which everything else would be futile or impossible. This opinion is held more or less openly, but in any case consistently. The other side sees growth as the basic evil to be eliminated. No growth or postgrowth is, from this point of view, the condition for a good life.

Second, neither of these positions asks on what empirical materiality growth is determined in the first place. They remain in the realm of almost ideological categories that they base on standard economic statistics that count all economic processes according to the money flow and without distinguishing their material basis. So, drivers of growth and postgrowth include the costs of disease and those environmental damages that are remediated (rather than avoided), while those costs that consumption and production are externalising are simply not counted. Therefore, they are not paid for in the supply chain and as respects the environment and human values, and do not appear anywhere. In other words, given the structure of this discussion, we may well be "growing poor" without noticing, and realising it too late.

Functional elites take a different approach. Such elites are understood here to be groups and people who are interested in operational, concrete action, who have means and roles in academic, corporate or public communities to pursue what they think fits the purpose. They can be found in all areas of politics. They explicitly are concerned about the operative side of "good governance" and they ask what figures, calculations, designs and procedures can be used to implement politically defined goals. For them, a goal is not already attractive and worth pursuing when it purely politically depicts a distant vision, but when its promoters have taken the trouble to operationalise that goal in concrete terms and timelines. They experience the relative remoteness of most political discussions from operationally thought-out practice as a devaluation of their work. They are not naively apolitical in this, but they stay out of what they see as political jockeying for position and patterns of power.

It is not surprising that it is precisely parts of the functional elites who look behind the categorical conversations. But it is surprising that they do so on the topic of growth/no growth, because for decades the consensus in economic discourse was that the practice of externalising costs (and risks as well as benefits) should not be discussed. In this context the functional elites in financial and manufacturing industries are not known for any will to transform the free market economy and its basic principles. Now, however, eminent parts of the financial industry and leading manufacturing companies are increasingly experimenting with calculating both full cost and full benefits that would internalise those so far social and environmental costs and benefits. This is not a Germany-alone conversation. From the outset "sustainable finance" has strong European links via upcoming EU regulations of the so-called "Fit-for-55" package (alluding to climate goals) and the next stage of requirements for corporate disclosure and green investments.

Third, the asymmetric distribution of power and solution thinking. Dialogue on how to share the burden would bring together young people who would be nominated by top leaders and mainstream executives – so that people who are at the end of their professional careers, but still in charge, would hand over responsibility to those who will spend most of their professional and personal lives cleaning up the carbon mess and biodiversity disaster. The distribution of decision power and the burden of delivery are asymmetric. A transfer of trust and confidence in individuals should be at the heart of a transformative strategy. So far, there are few explicit practical examples of this, carried out by the Council for Sustainable Development (sustainable futures) and the Michael Otto Environmental Foundation (dialogue on forest strategies).

Fourth, activist groups are raising public awareness by imagining collapse pictures, such as overheating the Earth, the planet's biological depletion. Such arguments are usually used to suggest and justify high ambitious policy targets. As they hardly achieve the immediate hoped-for consequences, protagonists pursue their mission even more resolutely as a kind of an absolute and robust "*j'accuse*". In addition, legitimate "green" concerns about the environment are often bitterly opposed to red concerns on social justice. Some fear the end of the world, others the end of the month (and their household money).

The opposite position, which dominates the political thinking in Germany, is also aware of the urgency of the issue and its scientifically based rationale. It tries to build on current policy-making and the positive experience of far-reaching change that has already been made, such as German-German reunification. It relies on the skills and competencies to find solutions, and less on shock messages. One finds this enlightened mindset among politicians of all democratic parties. But that doesn't mean it's fully catching on. While climate protection and sustainability are recognised as a political necessity by ratio, they are simultaneously felt and perceived as an imposition on democracy and those in charge. This tension creates the latent state of limbo in which the political sustainability discourse finds itself. Thus, the position prefers designing processes, even if they are slow and insignificant.

Both positions (not to mention the numerous slightly different variants) are gaining support, but proponents rarely engage each other in a meaningful conversation on long term strategies. This prevents more from happening. In particular, data-based and knowledge-driven approaches to sustainability and circular economy could tell the story of a desirable and achievable future.

Fifth, for industrialised Western societies, the leading social thinkers Ulrich Beck (1992) and Zygmunt Bauman (2006) describe fear as a general and widespread, deep sociological phenomenon. In German debates on climate and sustainability, Angst indeed plays a certain role, regardless of whether threats are material or whether fear is merely a feeling.

This type of collective angst is generated by multiple factors and attitudes. In some way or another all of them have to do with how reality is perceived: (a) Society appears overly complex, with complexity increasing. One feels constantly dependent on "the world" and how this dependency affects one's own life is hardly foreseeable and a source of insecurity. People cling to the old and clumsy narratives of growth that was supposed to be infinite, of globalisation that was supposed to create prosperity everywhere, of the megatrends of consumption that were supposed to enable freedom and self-esteem. And they are disappointed time and time again. Infinite growth is not possible in a finite world, globalisation divides more than it unites, consumption becomes a burden. Disappointment unsettles and scares people. However, the real problem is not angst itself, but the false expectations that create one's own weakness and subordinate position. Unfortunately, this fear shows a tendency to become empowered by its very existence. (b) The seemingly infinite recycling capacity of fear may only be overcome by using and defending complexity as part of the solution. Which it is, of course.

The first reflex of the ordinary politician, however, is to do the opposite and to promise people that he will manage complexity by reducing it to what seems practical. Even if he succeeds in doing this, he in turn unwillingly reinforces the aversion and may generate fear once again. (c) People feel increasingly threatened by permanently proclaimed multiple crises, be they terror, economic downturn, data surveillance, pandemics, environment, climate. More often than not threats are neither really tangible nor do they appear to be avertable. For instance, the political concept of multilateralism – although entirely correct as plain realism and needing further strengthening – is difficult to "sell". In terms of those fears multilateral collaboration seems to add fuel to the fire because it positively focusses on precisely those dependencies and uncertainties that are seen as part of the problem.

In conclusion, the actual conversation certainly is much richer than the structure presented here. However, "seeing through" the curtain of what is talked about in endless rounds and chats, one may discern some patterns: fear versus realism, lies versus truths, abstract threats versus concrete entry points of action, being victim to an anonymous loss of control versus sustainable action for self-esteem. US President John F. Kennedy in a speech at Yale University once put it like this:

> The great enemy of truth is very often not the lie – deliberate, contrived and dishonest – but the myth – persistent, persuasive, and unrealistic. Too often we hold fast to the cliches of our forebears. We subject all facts to a prefabricated set of interpretations. We enjoy the comfort of opinion without the discomfort of thought.
>
> (Kennedy 1962)

2.3 Change is happening

Germany is a society that is constantly changing, against of the odds and against what conservative politicians are trying to tell the people. It is constantly revisiting its history and finding new answers to the future.

This makes Germany different compared to other modern nations, and this is due to its recent history. As MacGregor (2014) put it, defeat in World War I, the collapse of the (first democratic) Weimar Republic and the murderous criminality of the Third Reich have made a reassuring coherent reading of the German history impossible. Instead, he observes, the sense of history, as it cannot be repaired, has constantly been revisited. The German political leitmotiv *Nie Wieder* (never again) is foremost a "never forget", and therefore a concept for the future. Even more so after the Wall came down and the two Germanys found themselves reunited in 1990, the German thinking about the national history is intertwined with a thinking about ways to a responsible and empathic future. Here, change is a programme.

This programme is not something one can find publicly announced and permanently talked about. It rather is an inherent feature of German politics. It is not always apparent, and there are many examples of nonsensical insistence on old ideas, of expensive blockades and of attempts to turn things backwards. But the experience of German reunification, as well as of flexible and successful steering through the economic and financial crisis of 2007/8, the defence of the European Union against populist breakaway movements, and the pandemic speak a different language. Political necessity is recognised, even if it is sometimes perceived as a democratic imposition. This is the background against which the German Sustainable Development Strategy, GSDS, must be understood.

The GSDS does not expose a formal theory of change. People did not invent a theory of change before they tear down the Berlin Wall. Experience has it that change may take surprising ways. Best practise activities and the (re-)telling of stories of success and failure are the essence a culture of sustainability. The rule of law, a strong social security scheme, and a clear federal constitution provide the safe operating room for public engagement in transformative change. In addition, Germany's parliamentary-representative system favours governments consisting of more than one party, and so, favours consensus-driven politics that, on top, have to make sure, the 16 states (*Länder*) that are forming the federal republic are heard and have their full executive, legislative and judicial power.

In Germany, Covid-19, climate-induced weather anomalies and biodiversity (insects) losses provoke a renewed appetite for sustainable transformation and collaboration. Any attempt to think systemically and to think outside the box needs specific tools, promotors and governance.

In the last 20 years, Germany produced more practical experience, goals, instruments and agreements on sustainability than in the entire period since the start of modern environmentalism with the first UN Conference on the Environment of 1972 in Stockholm. Those who begin to do things differently leave behind many routines and may sometimes be criticized for that. But experience teaches that getting started gives you courage. Never before have so many people been committed to not throwing away food, to recycling waste, to eating healthily, to campaigning for effective climate protection and to taking to the streets for an intact environment. Never before have such large parts of the business community recognised sustainability as a paradigm shift. Never before have politicians set more ambitious goals.

Most remarkable is the change in the culture of public consciousness. For some time now, the majority of citizens, buyers and consumers have been taking it upon themselves to act in a socially and ecologically responsible manner. The number and market power of sustainable companies have also been increasing. Sustainability management provides the company with more efficiency, a better reputation, more satisfied employees and higher qualification of young employees. There are also advantages in recognising early on where the next innovations are. In the end, this can even increase profits.

So who wouldn't want to be part of the current success model of sustainability? The times are over when sustainability could be spitefully ignored or benignly postponed until tomorrow.

While there are sound targets, programmes and plans galore, there is also delayed action, inaction and idleness. One needs to see both, and to discuss both, and more dispute about honest and truthful statements is necessary. Because today, even those who have not earned it are talking about sustainability: the German company Wirecard, which was embroiled in an accounting scandal and insolvency, professed its commitment to sustainability and publicly exhibited its corporate responsibility; the leading companies in the German meat industry are out with professional sustainability reports that stand in stark contrast to the poor working conditions that Covid-19 infections made publicly known. It is too easy to pretend that an entity is acting sustainably.

2.4 2022: Is Germany finding a way out of the stalemate?

Commitment to goals for sustainable development and climate may be on time, but action is definitely not. Pretty much all countries in the world face the paradigmatic picture of the valley floor (as shown in Figure 1.1) where action does not match commitments. This is raising interesting conversations in Germany on governance (market powers versus regulation) and on the mindset of a committed public (readiness for even hard action versus the attitude of "wash my fur but don't wet me").

Following 16 years of conservative rule, in Decembe 2021, a "traffic light" coalition assumed office, made up of the social-democratic, the green and the liberal party (red, green, yellow). Sustainability is on the forefront of their narrative, highlighting their respective political background (freedom, market autonomy, social responsibility, innovation and technical solutions, green growth) and almost without explicit reference to the German sustainability strategy. Consequently, the term sustainability appears thirty-five times in the coalition agreement, but only once is there a reference to the German sustainability strategy.

The "traffic light" coalition sees Germany in a phase of general repositioning. That accounts for foreign policy, agriculture and meat industries, mobility and urbanism, and, of course, climate change policies. An immediate stocktaking of where Germany stands in terms of greenhouse gases, as of January 2022, shows a drastic backlog. Climate targets for 2021 will not be met. Delivery for 2022 and the subsequent years are unlikely. As a result, the government claims it needs to triple the pace of emission reductions to meet 2030 targets, and thus achieve significantly more in less time. A solar acceleration package aims to make farmland available for solar, make solar mandatory for new commercial buildings, and commonplace for private homes. Onshore wind will be spurred by reducing minimum distances from military infrastructure and settlements. The renewable energy surcharge will be removed from consumer prices and transferred to the state budget to control rising electricity prices. Subsidies will be used to accelerate the decarbonization of industry. Another goal is for 50% of heating demand to be met by climate-neutral heat by 2030. Domestic production of hydrogen from renewable electricity will be greatly increased and imports of green hydrogen will be expanded. These first announcements have been well received and they clearly send different signals than before. However, there are also concerns and reservations about an absolute green growth (technology) paradigm being implemented in a siloed-in way. Conflicts of objectives remain, e.g. between biodiversity and wind turbines, renewable energies and land use, use and protection of forest ecosystems, or with regard to the ability to deal with rising prices. The fear is that they will simply be passed over instead of seriously mediating conflict solutions. This reflects some broader questions about managing transformation processes.

First, the focus on administration. Of course, administrations in general have to do their share in implementing of sustainable development goals. Globally, this is a given consensus. Therefore, national governments are the main contact and therefore institutions, regulation and administrative action are in the foreground. To what extent this focus is sufficient and whether the national level needs to be complemented by multinational and multilateral arrangements poses a key question that is yet unanswered. As in the German example, the focus on (inner-) administrative procedures and executive programmes (deliberately) steers the interest away from politics and the positioning of political parties – in the assumption that politics only disturbs and has done its part anyway, by having defined the goals. The "rest", meaning the implementation, is a task for the executive branch alone. It is a focus on administration, proposed by administration and supervised by administration. This attitude can lead to a depoliticisation of the issue, which would lead us down a wrong track, because most political decisions concern what people care about most, and that is implementation. At the policy level, this focus also comes at a high price, fostering a habit among policymakers to simply "forget" sustainable development strategies or to view these strategies as bureaucratic monsters that keep valuable staff from doing real work while they write reports and haggle over indicators.

Second, the lack of delivery unsettles the political actors. The bottom seems to them like the threat of new rules of the game for their careers and the patterns of power. For until now, education and career paths were at least halfway calculable. Rules of the game appeared to be known and accepted. The path to the 2030/2050 goals change that. The material environment can no longer simply be planned as before. Rather, there are principal uncertainties, coincidences and unknown interactions, and sudden events such as floods, landslides, droughts and infestations of pests. Whether it is genetic drift, carbon content in soils, mutations of insects and viruses, or methane deposits in permafrost, this dynamic is a new challenge for actors in politics and economics. It creates a paradoxical relationship between knowledge and certainty: until now, a great deal of knowledge has always generated a great deal of certainty. People have been used to this for centuries. Now, however, the amount of data and knowledge about the climate and unsustainable global trends, as well as about new solution strategies, is constantly increasing. But the more data available, the more certainties dwindle. They are replaced by probabilities and a peculiar coupling of action and certainty.

Political actors in Germany are often used to the idea that everything is interchangeable: people, parties, relationships, interests, money, power, careers, jobs and entrepreneurs. In this understanding, politics is a barter that is good when it is organised in a comprehensible, honest and useful way. For example, the idea of the cultivated landscape suggests that sowing, fertilising and harvesting can be controlled at any time and are mutually exchangeable variables. But now a new factor comes into play. There is an inherent power of the material. Ecosystem responses create new niches for evolution. This is hard to predict with certainty. This nature cannot be replaced by technology and capital.

In the persisting old world, politics "play for time" and may be tempted to sit out problems. Democracy means *rule for time*, and its main characteristic is that parliamentarians get elected for a certain period of time. That of course allows for such time-consuming measures. But now issues of sustainability policies enter the realm of politics, emphasizing the intergenerative motifs to the ethics of justice and accountability. How do they fit together? Postponing measures no longer creates a breathing space, but only brings back the material effects of global warming and destruction of nature all the harder. Postponement is aggravation. Are politicians aware of this? And are administrative "owners" of sustainable development strategies aware that they have a shortcoming, and that they lose control?

Normally, losing control is an issue for all executive systems. It is exactly what should not happen and needs to be avoided by all means. When something does happen that can be regarded as a loss of control, the mainstream declares it a crisis. The word suggests that the situation will pass again. In a way it stands for the mindset that one has to get through the issues now, and that maybe exhausting and impositions are unavoidable, but afterwards everything will be just like before.

But this political conception misses the point that Germany's key economic, ecologic and social figures, and its position in the larger world, suggest major internal changes in terms of demographics, climate, immigration, digitalisation, financial spending and debt. New orders and rules of the game are needed. Sustainability strategies are the right way to approach change in a manageable and transparent way. For politicians a sustainability strategy holds some advantages. First, it suggests that one has "things under control" and that there are ideas for the future. Second, a strategy can connect to basic mainstream concepts such as liberalism, conservatism, social and democratic policies, as well as politics for greening societies. This message is attractive to policymakers in any country, but will it be received? Without it, fragmentary individual actions and dependencies threaten to drag politics into a vortex that is inevitably downward and undemocratic. But is this also seen?

Obviously, there is a huge amount of grey area here. And that, exactly, is why governance and its mutual link to mindsets and preparedness of actors are making a difference, as we will see in the following chapters. They can encourage sustainable development and be part of a solution. Or they can stall any improvement and even trigger turns for the worse.

References and recommended reading

Bachmann, Günther (2017) Deutschland muss Radikalität lernen. In: Wolfgang Ischinger und Dirk Messner (Hrgs.) *Deutschlands Neue Verantwortung: Die Zukunft der deutschen und europäischen Außen-, Entwicklungs- und Sicherheitspolitik*. Berlin: Ullstein 431S, 360f.

Bachmann, Günther (2020) *Die Stunde der Politik: Essay über Nachhaltigkeit, Utopien und Gestaltungsräume*. München: Oekom Verlag.

Bauman, Zygmunt (2006) *Liquid Fear*. London: Wiley.

Beck, Ulrich (1992) *Risk Society: Towards a New Modernity*. London: Sage; German original published in 1986 by Beck in Munich.

Brundtland, Gro Harlem (1987) *Our Common Future, from One Earth to One World*, www.un-documents.net/ocf-ov.htm#1.2; https://

Credit Swiss (2021) *Global Wealth Databook 2021*, www.credit-suisse.com/media/assets/corporate/docs/about-us/research/publications/global-wealth-databook-2021.pdf

DESTATIS, German Statistics Office, website, www.destatis.de/EN/Home/_node.html

Federal Government of Germany (2021) *German Sustainable Development Strategy Update 2021 – Summary*, www.bundesregierung.de/resource/blob/992814/1942596/dff76fd102a1e14aafbe53de4f20a462/deutsche-nachhaltigkeitsstrategie-2021-kurzfassung-englisch-bf-download-bpa-data.pdf?download=1

GIZ (2018) *Germany in the Eyes of the World*, www.giz.de/en/worldwide/63559.html

Kennedy, John F. (1962) Commencement Address at Yale University, June 11, 1962, https://www.jfklibrary.org/archives/other-resources/john-f-kennedy-speeches/yale-university-19620611

MacGregor, Neil (2014) *Germany. Memories of a Nation*. London: Allen Lane.

Mauch, Christof (ed.) (2004) *Nature in Germany History (Studies in German History, vol. 1)*. Oxford and New York: Berghahn Books.

Welzer, Harald (2020) Alles *könnte anders sein: Eine Gesellschaftsutopie für freie Menschen*. Frankfurt am Main: Fischer Verlag.

3 Roots

The background of today's sustainability practise

3.1 Roots of the future

The past is never quite past, and historical roots are never just history. Roots to the past are connecting us to the future. Sedimented events, relics and personal stories make us rethink about how present and past cling together and what we make of it. They grow into the future, for the better or worse. With the idea of roots we are constructing what was considered important in the past and what we still consider important today, with today's experiences and in the context of current politics. This is not worrying and certainly nothing sensational, because it applies in quite the same way to our personal life and any person can test this principle if one wants to.

The roots in the past are constantly reconstructed. The reconstruction reinterprets all the relics, some of which are lost or overshadowed by others over time; others are highlighted, illuminated and placed in a larger framework. As has been often and impressively demonstrated, this mixture of losing and finding (with all its shades of grey) is a cultural phenomenon (Assmann 2011).

As far as modern political thought is concerned, all of its roots ultimately lead back to the ideas of the Enlightenment, which is the reason why serious political thought still attracts so many people. Enlightened politics puts the individual first: the enlightened person is rational and reasonable and confirms his/her place in society through work and functionality. Reason, freedom, and the assumption of unlimited educability of man through scientific knowledge become measures of progress. It is in this ideal that people place their hope for their own advancement and for the recognition of their learned skills. And it is precisely here where the root of the most serious defect that all modern political thought lies.

The ideal has a dark flip side. It is the almost total blindness to the environment and the natural life support system. The Enlightenment recognises work only as a relationship of people to people, but not as the human appropriation of natural resources. In this respect, enlightened reason only regulates the relationship of the members of society to each other (which is a formidable task and will remain a challenge at all times), but not the mutual interaction between man and nature, which turns out to be an even greater task.

DOI: 10.4324/9781003242024-3

The concept of sustainability tries to respond to this and to fill the gap. That is why it is misleading and even counterproductive to relate the concept of sustainability exclusively to the environment per se, quasi as the better protection of nature. However, many do so in the belief that they are particularly timely because global warming and species loss require more ambitious policies (which is indeed the case) and a purely environmental approach to sustainability (often referred to as deep sustainability) would suffice. This belief is false.

The real strength of the concept of sustainability comes from its breadth because with this it is tied to the Enlightenment and its appeal to sense and reason, to freedom and commitment. Ulrich Grober (2012), for example, traces it back to a whole range of historical texts, from Francis of Assisi, Nicholas of Cusa, Spinoza, Lenné, Goethe and Novalis to the scholars of Tharandt, the old German centre of excellence for forestry, and the mining director Carl von Carlowitz in nearby Freiberg, as well as the great explorer and discoverer Alexander von Humboldt. According to Grober (2013) the term sustainability has a career trajectory from Carlowitz's Freiberg to modern UN policy.

Unlike Europe and Germany, Alexander von Humboldt is still very well known in the countries he visited on his expeditions: Colombia, Cuba, Ecuador, Peru, Mexico and Venezuela, but also in whole of Latin America. In Latin America he is deemed the mastermind for a holistic way of looking at nature and societies from a globally responsibility perspective. His example teaches us to be concerned with how the ecology of life interrelates species, their spatial distribution, geography and mankind. Humboldt linked culture, territory, sustainability and people long before the hegemonic eurocentrism began to dominate Latin American thinking. And, thus, he established what later would help develop the rich thoughts of Latin American environmentalism as described by the long-time coordinator of UNEP's Environmental Training Network for Latin America and the Caribbean, Enrique Leff (2010).

It is a fair assumption that every culture has developed some way to reflect its relationship between humans and nature. In the German case specific domestic roots complement those from international contexts. Tellingly, all the roots have a common feature. The protagonists took unusual paths, had to think outside the box, and not infrequently transcended the usual procedures and communication channels. They went off script. Otherwise and in terms of content, mission and purpose the roots show variations and differences. It is hard to track down unique and solitary impacts in the way they changed the course of life. Very generally, attributing distinct or even deterministic causes for either societal collapse or survival in a sustainable modality must always remain doubtful. In his book on how very different societies failed or succeeded Jared Diamond (2005) speaks out against determinism when he insists that "environmental problems do constrain human societies, but the societies' responses also make a difference".

Tracing roots can also raise new questions that confront the present and the future. Take the history of industrial buildings, for example. In the 19th century, urban development experienced an unprecedented boom: Manchester and

Liverpool, Berlin and Leipzig – all recorded an annual population increase of more than 10 percent, echoing the increase in industrial production. Manchester was rightly called a shock city, referring to its poor working and living conditions. It was no big different in the other booming "tiger" cities. Noise, dirt and stench displaced urban forms of city life. It is an interesting question why we see so many differences today in the ancient industrial culture and its building relics (Osterhammel 2015, 2009), and what this means for our understanding of the third industrial revolution?

Understanding the roots of the ideal of a decent life within the carrying capacity of the natural environment can show what a difference society's responses and the added value of protagonists, activists and thought leaders can make. Remembering and retelling roots is more than a bookkeeper's effort of historical accounting. It has a meaning for today. Invoking historical root lines establishes and strengthens today's "we". The invocation of primal ground, ancestors and lines of tradition legitimises institutions, policies and demands, but also self-reflection and ethical standards. The least that can be said is that the future builds on an understood past, as does sustainability thinking on the Enlightenment. Roots are catalyst for creativity.

3.2 Germany's "Deep Carlowitz" and today's issues

If you want to get to the bottom of the origins of the sustainability idea in Germany, you can't get around Hans Carl von Carlowitz, who lived and worked in Saxony 300 years ago. His story was forgotten in Germany for just as long. Today, it is often retold from new perspectives.

Carlowitz was what today would be called an economic minister of a developmental tiger state, an emerging economy, or a chief financial officer in a major corporation. His CEO was the Elector of Saxony Augustus the Strong, who paid him to secure and increase the princely income.

In Carlowitz's day, the entire economy revolved around forests and mining, the latter depending on wood. Since the 15th century, mining in Saxony was reserved for the sovereign. The upstream and downstream technologies and trades, on the other hand, were the responsibility of the citizens. The relationship between the core business (mining, princely guaranteed) and competition in transformation is the source of the "Saxon ingenuity" with which the craftsmen and engineers made their Saxony of that time the heartland of mechanisation and technology. Thus, any economic growth depended on mining, and mining in turn depended on wood: to develop the underground shafts and to smelt the ore. In the Saxon forest, overexploitation prevailed. There was talk of a devastating timber shortage. This was not only the case in Saxony, but in many places in Europe. On his own estate in Arnsdorf, Carlowitz experienced extremely dry summers and storms and consequently the infestation of his forests with bark beetles. That was drastic enough, but even more dramatic to him seemed the impasse toward which – translated into today's terms – the Saxon Mining Innovation Cluster was heading. The impasse threatened the existence of one of the most important economic regions in Europe at the time.

To understand the scope of this insight, one must imagine the attitude to life at that time. The Thirty Years' War (1618–1648) was probably as vivid in the memory as World War II is today. And it had raged cruelly, especially in central Germany. Large armies had devastated the country by their mere presence – even when they were not in battle. During Carlowitz's lifetime, the plague epidemic, the Black Death, repeatedly raged through the country. In many regions, two-thirds of all people died of the plague. How must it have affected the people when their prosperity and the basis of their lives were once again threatened?

For the prince, mining was the decisive source of wealth, luxury and power, even if this meant blatant overexploitation of the resource wood. The weal and woe of the country hung by a thread, or rather on the timber. Did the fundamental timber crisis shake those responsible at the time in a similar way as the banking and financial crisis did in 2007/8? It was against this background that von Carlowitz wrote his work *Sylvicultura oeconomica* in 1713 (Carlowitz and Hamberger 2013). This was not meant to present nice textbook solutions. In style and message it rather was a mission statement, a call, an appeal and political demand.

In essence, he was concerned that no more wood be taken than will grow back; consequently, that those responsible must know clearly and comprehensibly how much wood is growing and what quantity can be harvested in an economically viable manner. Carlowitz called for an end to wild overexploitation and advocated the introduction of a proper count of the trees. Harvesting should be limited to the amount that would grow back, an ancient version of the limits to growth. Carlowitz established rules for forestry in order to adapt it to the forest ecosystem. However, all of this never materialised to the full extent. Soon, the era of fossil fuels was on the horizon. The industrial development of coal as an energy source finally relieved the forest of its function for the smelting of ores. Carlowitz's ideas fell into public oblivion for centuries.

The critical reconstruction of Carlowitz's thoughts always reassembles what constitutes cultural memory. Therefore, Carlowitz's message remains a contemporary category as the Saxon Carl von Carlowitz Society, a civic network, underlines with many actions.

Carlowitz was not alone in Saxony in his time, although networks and cooperation in today's sense did not exist then, of course. In many hot spots in Europe, foresters were thinking about the future of forests. Wood was like oil or gas in the 17th and 18th centuries, and its use was sometimes rigorously regulated. At that time, cultivation of the landscape was imagined as subjugation of nature. Beauty was usefulness, and useful was that which increased wealth. Carlowitz was more radical in this context, because he was concerned not only with numerical control of the number of trees in the forest, but with precaution (sufficiency, frugality) as limit to growth, and with innovation (substitution of wood) as a transformative approach to wellbeing. Today, these are the inconvenient truths of ecology, as the scientific doyen of German ecology, Wolfgang Haber, called them in the first Carl von Carlowitz Lecture of the Sustainability Council (Haber 2010).

Anyone who remembers Carlowitz today must therefore go beyond him. In doing so, questions arise about which one would be only too happy to know what Carlowitz would have advised: the forest as carbon sink, as climate saviour? Optimising productivity or leaving the forest as wilderness? What about the tipping point of the Amazon's self-deforestation? What about the disappearance of spruce from Germany's forests? Whether it's smart to send half of Germany's timber harvest through Germany's roughly 12 million chimneys? The forest ecosystem is already at its limit and can hardly help to mitigate climate-related heating. Certainly, the tropical rainforests are not an infinitely resilient CO2 sink, and it is also certain that the Central European forest will rebuild itself and adapt to the climate if it is left to do so. It is reasonable to assume that a today's Carlowitz would become restless. Possibly he would be at the forefront of those demanding limits to growth and advocating innovative high-tech solutions with wood: Would he wear shirts made of cellulose, revolutionise building construction with wood, make the bioeconomy a leading industry?

Of course, we won't get any answers to these questions. But the answers would only be of secondary importance. What is of primary importance is that we learn to ask these questions, to think in terms of what is possible, and to create images of the future. I have only scratched the edge of what is conceivable. By way of example, wood can in fact already replace steel in building construction and be used as a liquid material.

Carlowitz had the right narrative in his day, and he lacked transformation. Today, there is a lot of transformation, but the grand narrative seems to be missing.

3.3 The momentum of Gro Harlem Brundtland and her group

In 1983, the UN General Assembly appointed Gro Harlem Brundtland, then the acting prime minister of Norway, to head a world commission. At that time, world commissions were a frequently used political instrument to deal with new issues. The one headed by Gro Harlem Brundtland would soon prove to be the most famous and influential commission in a world that was highly divided.

When Brundtland set out to put the concept of sustainability on the world's political agenda, she changed the order of the world. The rifts between East and West, North and South, South and South were deep. They made any thought of a world-historical turning point seem absurd. Latin America suffered from the militarised politics of post-Pinochet Chile. As a European, it may be hard to believe today that terror dictatorships had been shaken off in Portugal, Greece and Spain only a short (historical) time before. Berlin was still divided by the wall that everyone thought would last for a long time. In the US, President Ronald Reagan shaped the 1980s by, among other things, scaling back the environmental commitment of his predecessor Jimmy Carter. Many countries in the Global South had freed themselves from colonialism after long struggles. They now claimed economic liberation and wanted a fairer world economic order. Commodity countries demanded sovereign access to global markets and independent pricing. Even today, 90 percent of oil exporting companies are state-owned by their respective countries. Consequently, there was a very fundamental dispute over how to regulate global trade and foreign investments, or the right to food.

This could have gone on for a long time had it not been for the bottleneck of ecology. Economic interests encountered limits due to environmental pollution and overexploitation. The insight that all economy should ultimately always and everywhere be aware of the vulnerable human-natural interface was in the air, but not yet a political reality. The issues around climate change were not widely understood in the 1980s. Losses of biotopes and species were known and complained about, but it was not as well documented for a long time as they are today. However, what was known and could be known about pollutants, the ozone layer, hunger and malnutrition, poaching and the negative effects of urbanisation was long enough to draw political consequences and demand a different model of development. But how to do that practically was either unclear or politically contentious.

This puzzle was proof of how deeply frustrated the international community was. The community of nations felt inadequate and ill positioned. In the context of the dispute over economics and sovereignty, environmental concerns had not been a factor at all. This increased the complexity of the North-South issues. If the Commission was to make a difference, it had to be different from anything that had been used to. Brundtland's impulse had method. It began with the composition of the Commission and extended into the practical nature of its work.

Gro Harlem Brundtland first secured high-profile comrades-in-arms. William Ruckelshaus, one of the first US environmental politicians and one of the upright democrats who uncovered the Watergate scandal, came from the United States. To secure Germany's support, Brundtland approached Helmut Kohl, then Chancellor of Germany. Helmut Kohl is generally not known for environmental ambition and clearly wanted to keep the issue away from himself as much as possible. Predictably, he asked his predecessor and former leader of similar world commissions, Willy Brandt, to make a proposal. In this way, he secured high-level representation and, from his point of view, a safe distance from his government policy at the same time. Volker Hauff, former German Research Minister and strategist for innovation and modernisation of the German economy, became a member of the World Commission. Jim McNeil from Canada took over as executive director, seamlessly continuing the tradition of progressive policy-making that Canada stood for at the time. Brundtland made former Sudanese Foreign Minister Mansour Khalid her deputy. The commission was joined by 16 other members from around the world, including (for the first time in this framework) China.

The commission developed a new modus operandi. Opening up its negotiations to citizens and lay people it established an early form of an open method of deliberative solution search. It set itself apart from expertocracy. It took members weeks to speak with affected communities and representatives of indigenous people. This led to two-week – partly public – meetings in Jakarta, Oslo, São Paulo and Brasilia, Vancouver and other places in Canada, Harare, Nairobi, Moscow and Tokyo. This exercise in listening and collecting facts was a clear rejection of the usual forms of negotiation, which mainly make judgements on the basis of pre-set opinions. The atmosphere of such meetings developed from personal insights and emotions. In addition, subcommissions and specialists worked on sub-topics.

It would have been normal if the Brundtland Commission had coined an epoch term (another XY-ism). But it did not. Rather, it proposed a political process in which it refrained to perpetuate itself (again a move that was not self-evident). As Volker Hauff pointing out it makes a difference whether you want to be right and therefore put yourself in the controlling position, or whether you want to keep right and therefore empower others and engage them with your principal idea. Thus, Brundtland placed the follow-up process in the hands of the member states of the United Nations. To this end, she gave them the political concept of sustainable development and an idea of how this concept would best be processed within existing political structures. Goals for ecology, economy and social affairs should be worked out and democratically legitimised, they should be measurable and it should be possible to verify them. Public transparency and distributed responsibility should help to manage the commons. Analysis should not stand by itself, but be linked to imagination, describing possibilities and opportunities.

The German government and non-governmental communities received Brundtland's message and the definition of sustainable development with a bias. In the original version, the central concept in this definition is the word "without compromising". It links the present and the future with an offensive activity. However, the German Zeitgeist translated the Brundtland text with the defensive word "risking". In German, the term risk suggests a defensive attitude, which is basically a biased translation. This bias might be explained by circumstances at that time is which was characterised by millions of victims of the African and the Ethiopian famines, the chemical disaster of Bhopal, the poisoning of the Rhine by the Sandoz company, Mexico's gas explosions with thousands of deaths and the meltdown in Chernobyl. Half of Europe was radioactively contaminated, and there was great fear for the future. In 1986, a freight train carrying milk powder contaminated by Chernobyl wandered through Germany for days because no one knew how or where to dispose of the material. Risk and responsibility appeared in a new dimension. The commission observed the collapse of the ozone layer. Sociologist Ulrich Beck hit the nail on the head with his book *Risk Society*. According to Beck, the new, transboundary environmental problems and, in particular, the completely limitless consequences of accidents in the use of nuclear power make risk the dominant negative experience. The risks of nuclear power fatefully bind all futures to this technology and its eternally radiating waste, Beck said, which is why the risk society is a burden on the environment and democracy (Beck 1992).

Those events and that debate accompanied the Brundtland's deliberations. The German translation may have rightfully chosen the word "risking" as the crucial anchor term. As compelling as this might have been, it sets policies on a particular risk-aversive path. It suggested avoidance, restraint, and renunciation, but not so much making possible, exploring new paths, participating, creating commitment, and choice editing. Even today, the German concept of sustainability suffers from this.

In 1992, this Brundtland momentum led to the Earth Summit in Rio de Janeiro. "Rio" translated the Brundtland postulate into policy, and the Rio Agenda 21 triggered some sort of euphoria worldwide. After all, the Rio Agenda 21 brings about a climate agreement and agreements on biodiversity and combating desertification. The hole in the ozone layer is stopped. Agenda 21 calls on the countries of the world to each develop their own national sustainability strategies. This is a success. German Environment Minister Klaus Töpfer led the German Rio delegation and was a key driver of the negotiation process. Later on, as UN Under Secretary General and head of the UN Environment Program, Töpfer continues the Rio process and initiated additional impact.

Back in Germany, however, in the 1980s, Helmut Kohl's government refrained from presenting the Brundtland Report to the German Bundestag. While Japan, Canada and the Scandinavian countries supported the work of the Brundtland Commission with sums in the millions, Kohl's government left it at 60,000 marks. There was no parliamentary debate on the conclusions of the Brundtland Commission. This would be unthinkable today, when there are regular plenary debates on sustainability. But back then, things were different. Helmut Kohl neither paid attention to Volker Hauff's work in the Commission nor thanked him. Federal President Richard von Weizsäcker did, however, in a personal letter emphasising the importance and great merit of the Brundtland Commission's work. As the then-President and the then-Chancellor both are members of the Christian-democratic party this different response sheds a light on the rift in the German political class and on the fact that the leading conservative policy makers failed to understand the sustainability agenda and, as a consequence and with the exemption of Klaus Töpfer, political conservatism did never play a proactive role in establishing the new environmentalism. This discrepancy is still noticeable today.

It is not surprising that Helmut Kohl's cabinet, in the decades of 1980 and 1990, never discussed a sustainability strategy. Yet there has been no lack of intensive effort on the part of the Environment Ministry, lead first by Klaus Töpfer and later on by Angela Merkel. On her request the most important opinion leaders on ecology and society came up with high-quality preliminary work. The Wuppertal Institute brought the concept for a "Sustainable Germany" into the discussion. The Federal Environment Agency produced a comprehensive study entitled "Sustainable Germany". Parliament promoted the development of a sustainability strategy with an Enquête Commission. All to no avail: drafts for a national sustainability policy never even made it to cabinet level.

In 2007, on the occasion of the twentieth anniversary of their report, Volker Hauff, Gro Harlem Brundtland and Jim McNeil took stock in New York on the sidelines of the United Nations General Assembly. For Gro Harlem Brundtland, the glass was full. She emphasised that every beginning of a great achievement is at first on the edge of madness or at least on the edge of what is usually called normal. Volker Hauff drew a more critical balance and called for a necessary update of sustainability thinking. He said that the Commission had not given enough consideration to the issue of governance. According to him, a new approach to governance should overcome silo thinking of entrenched rope teams, sectoral egoisms and the echo chambers of specialist policy. There is also a need to anchor the multistakeholder approach institutionally in sustainability strategies. Jim McNeil warned that the Global South is still and increasingly being left behind and can hardly share in the benefits of good sustainability policy as there are no sound policies and tools facilitating mutual learning.

This critical view of missing links and deficits was the state of affairs for a long time, a time that can be seen as two lost decades.

3.4 Rio: an almost lost impact

If you can't think the impossible, you don't know the possible. That is the backbone of the Rio-story. It is a story about both positive sides and a decades-long paralysis. For the first time, "Rio" set the principles of environmental protection and economic growth (development) in relation to each other. For the first time, the G77 group of 134 developing countries from Asia/Pacific, Africa and Latin America, including China and India, and the OECD countries, accepted the basic view that the planet cannot be preserved for future generations unless more responsibility is taken and unless the idea of sustainability becomes the global guiding principle. Rio directed people's hope toward what seemed a new historical epoch that would overcome the East-West divide and end an arms race which would save the world gigantic financial resources and free those resources for humanitarian purposes. There was a belief in an era of reason, progress and sustainability. This was an important source of strength, but at the same time it was an illusion, as we know today.

The globalised hardening of the North-South divide, neoliberal deregulation, and financial turbo-structures soon pushed the Rio hope to the side. Geopolitical conflicts, disintegrating states and poverty drives millions of people from their homes. The neoliberal variety of globalisation diverts political attention away from the common interests of all countries and toward particular interests. As devastating this is, it still could not stop first sustainability policies and first climate mitigation efforts from emerging.

The loss of a decade or two is only inadequately covered by an endless series of talkative conferences with the same appeals. The loss comes at a price, also in a deep economic sense. Growth rates still mask serious systemic failures and the explosion of hidden environmental and true costs that are still not priced in.

Environmental degradation, as a cluster risk, triggers migratory movements and conflicts over natural resources.

What we (can) know today about what causes unsustainable trends, and what are predictables and unpredictables, we owe to research. Fortunately, in 1988, United Nations environmental policymakers, together with the World Meteorological Organisation, established the Intergovernmental Panel on Climate Change, a centre piece of improved governance: (a) the panel gives science a voice that cannot be ignored, and (b) it organises the political dialogue about its findings. It (c) also has an impact on science itself, because it facilitates determining the "state of knowledge" more quickly and efficiently than the usual procedures can do.

The real deficit today is not in understanding the past and the present, but in not understanding the future. For understanding the past and the present, scientific knowledge and data can provide an essential background. This is not equally true for understanding the future. Manoeuvring into the future depends on culture and an understanding of the commons. Many think that all you need to do to be fit for the future is to design goals, targets and timetables. Particularly radical ideas are characterised by the fact that they demand to cut the time until the goals are achieved. A competition for ever higher goals has arisen. It is almost a kind of mechanical and simplified accounting of the goals. Oversimplification is the flip side when goals are made absolute as policy instruments. Goals and targets are certainly needed, and ambitious one's in particular. But this is a necessary and not a sufficient condition. The sufficient condition is context, e.g. a comprehensive framework for sustainable development. This is exactly what was also lacking at the international level in the early 2000s, and what once again brought the Brundtland momentum to a near standstill.

In 2000, the then UN Secretary General, along with a small group of academic development economists, announced the Millennium Development Goals (MDGs). With the MDGs, they wanted to shape an era. Initially for 15 years, these goals were intended to symbolise and at the same time guide the relaunch of development cooperation into an age without hunger and poverty. That was the claim, but in the end the time of the MDGs ended rather silent. They did not continue the Rio discourse, and basically were a move backwards to the time before Brundtland. In the seven MDGs environmental concerns appear as an extraneous side issue. During the 15 MDG years, poverty could certainly be reduced, and some population groups advanced economically. It remains questionable, however, to what extent these successes are primarily due to the worldwide globalisation boost and, in particular, to the economic upswing in China. Moreover, the MDGs were limited in content to what was easily measurable in terms of conventional development economics. This strengthened the departmentalised silo thinking of the development community in the North and South.

The MDGs represent 15 lost years for sustainability. The MDGs once again cemented old-style development policies, including neoliberal economic concepts, including the thriving of fossil economies and economies built on extractive industries, global nature depletion, including a false model of consumption that drives people into obesity; soon, the number of obese people exceeds the magnitude of those many hundreds of millions who still suffer from hunger. All of this necessitates a new paradigm. But first, things got worse.

3.5 The mother of all failure, or a short glimpse of historical truth?

The debacle of the Copenhagen climate conference in 2009 is an icon of failure, people say. Indeed, in 2009, world leaders spectacularly failed to reach an agreement on climate protection. Nobody ever fails alone. Virtually the entire negotiating community – a few tens of thousands of representatives from governments and civil society, business and NGOs – had miscalculated the global shift of economic and political power that requires new solutions. Years after 2009, the metaphor "Copenhagen" still dominates international debates and mental maps. People speak of the curse of Copenhagen. An image went around the world: the most powerful heads of state in a small room, spontaneously cramped together at a table, apparently without a clue how to deal with the papers prepared by their diplomats that contained a lot of square brackets, showing large areas of disagreement. A few minutes later, they were to part without result.

The debacle contributes to the then widely shared impression that multilateralism is at an end and lives longer in people's minds than the problem itself. For a while, the picture gave the impression that cultural pessimism was the new realism. Faith in negotiated solutions falls low because it has previously been highly exaggerated. Even the UN Secretary General had publicly expected Copenhagen to seal a deal. In hindsight we know that all this were only inconsistent wishes and ideas of the participants and commentators in the run-up to Copenhagen. Moreover, the final negotiations of the 115 presidents and heads of government were so extraordinarily badly organised by the hosts and so opaque that many countries and observers switched from skepticism to suspicion and then to open rejection. That fuelled old reservations. The accusation directed at the industrialised countries is neocolonialism, which deprives the Global South of development opportunities and uses climate protection as well as the transformation to a green economy only as a pretext for a modern supremacy. A wave of discontent and mistrust resulted in non-decision-making.

Consequently, the Copenhagen conclusion, the so-called Copenhagen Accord, remained without official status. It is merely "noted" which among diplomats is considered the weakest category of treatment for a document. For international diplomacy, Copenhagen is the zero hour of trust, and the participants also no longer trusted their own abilities. The Copenhagen Accord is the most intensively prepared and highest stakes Non-Agreement of modern times. It is better called the Copenhagen Discord.

But the Copenhagen dispute also has another face. It offers a brief glimpse of historical truth. In this narrow space, a historical epoch ended, the climate policy à la Kyoto with its awkward construction of two camps. The Kyoto Protocol, an international treaty expanding the 1992 United Nations Framework Convention on Climate Change, divided the world into industrialised countries obligated to reduce their CO2 emissions and developing countries with no binding reduction targets. In the meantime, however, the material basis for climate policy has changed significantly since 1992. China, India and Brazil have emerged as extra-large carbon emitters and global economic centres at a pace that was unpredictable 15 years ago. And at a pace that was equally unpredictable, natural disasters have reached vulnerable countries. As a result, camps have multiplied. This regrouping is a truth that the preparations for Copenhagen have not really taken into account. Also, in Copenhagen clashed two mismatched governance principles. The climate negotiations, led by the signatories to the Climate Convention, are based on the consensus principle, giving an equal voice to the least developed countries and the states most affected by climate change. In the universal UN system, the large developing countries and emerging economies have procedural prerogatives that they zealously guard.

The Copenhagen Discord is still a bad omen for all multilateral climate policies. The Conference of climate treaty parties, held 2021 in Glasgow, had the Copenhagen curse literally painted on the walls of the negotiation room. It has become a mantra, although the Glasgow deliberations are by no means comparable to the 2009 situation. 2009 the world experienced a system breach, the 2021 Glasgow conference, with preparatory backing from major diplomatic efforts of the US President, the French President, the host country, Germany and the finance industry, is putting a system together.

Nevertheless, the omen from 2009 also had a negative effect on other negotiations. It was the worst of all possible for the twentieth anniversary of the Rio Earth Summit in 2012 (Rio+20), or more precisely for the start of those negotiations that were supposed to follow up on the Rio agenda and renew the Rio momentum. In the preparatory phase for Rio+20, discouraged diplomats introduced only timid issues, and those without much political dimension. It was as if they had succumbed to the Copenhagen curse. It can hardly be surprising, then, that discord, ill feelings, and silly cantankerousness immediately arose where rapprochement, curiosity, and respect were needed. The idea of a Green Economy was denounced as an attempt to subjugate nature once and for all to the capitalist profit calculus, alternatively as domination of the North over the South. This has nothing to do with the factual reality of the Green Economy, but it was a diplomatic rally. Technical standards, e.g. stewardship criteria for sustainable production, were denounced to only defend the dominance of the top dogs against imitators and economies that were catching up. Sustainable management was not seen as restructuring for all, but as a power play by protectionists. Others saw sustainable business as a Trojan horse of industrial lobbies that would sell noble sustainability to cheap profit interests. As for the governance issue, the air was no less poisoned. African states flatly refused to deal with it because they feared another conspiracy here. They thought the UN was trying to wrest from them the only seat of a UN institution in Africa (Kenya, United Nations Environmental Program). Paris and Geneva (other UN seats) were said to be sucking Kenia dry, leaving only an empty bureaucratic shell in Nairobi.

The spiral of mutual reproaches and fears for one's own position turned faster and faster and produced more and more whimsical phenomena. A second Copenhagen loomed. There seemed to be no way out. The lowest common denominator became microscopic. And even it met with reservations and veto positions.

3.6 Going off-script: the SDG momentum

Then something happens that no one expects. A proposal is put on the table that changes everything. It irritates. And it comes from an unusual direction. It is two women who bring it into the male-dominated diplomatic world. And it is a (economically) small country, Colombia, that is taking the proposal to the big players. The proposal thus comes from the Global South, not from the regions of the world that have been used to hegemonic politics for centuries. And it comes at an inopportune last-minute time.

Paula Caballero and Patrizia Londono speak for the Colombian Ministry of Foreign Affairs, Caballero as head of the Department of Economy, Social Affairs, Environment, Londono as acting Deputy Foreign Minister. The Colombian President and later Nobel Peace Laureate Juan Manuel Santos supports them. Nationally, Colombia itself is in a process of deep reconstruction and self-reflection. Bold thinking is needed to put an end to the civil war, kidnappings, terror and destructive violence of the North American drug markets. Caballero and Londono feel that the diplomacy with the smallest common denominator (and that would be almost zero) will not interest a broad public and does not get to the roots of the problems dominating the planet either. The future is not won this way. They feared diplomatic negotiations remain boring and stiff-necked in the paths of exactly those circumstances that led to the problems.

So they make a proposal that turns everything upside down. In mid-2011, Caballero proposes a new way at a preparatory conference that was supposed to close the basics and take care of only some details. Caballero offers a set of globally valid sustainability goals. Not downsizing the diplomatic bag was her solution, but drastically increasing it – not cutting corners, but designing for the future. What her idea offers is a future treaty on par with the Declaration of Human Rights. The initial response she got was: too much, too late, out of sync, a phantasm. Through September 2011, Caballero and Londono are talking informally with regional UN offices for Africa, Asia-Pacific, Central and Eastern Europe, Latin America and the Caribbean, Western Europe. The response is friendly but sceptical. But they were undeterred even by slightly pitying and patronising comments. They win Guatemala as their first supporter country. In November 2011, Colombia and Guatemala officially approach the UN through diplomatic channels. Arguing that the challenges are global and therefore need global action, they call for jointly developed benchmarks and guidelines. At the time, they were still talking about a post-2015 framework, expressing a reference to the MDG mechanism. But for post 2015 the two countries propose the adoption of Sustainable Development Goals. Broad political support initially fails to materialise. To everyone, SDG were a competition to the MDGs and their community, irritating and aberrant. To the mainstream SDGs seem to be just another half-baked and crazy idea, whose discussion would again take hours and hours before the idea would be shelved anyway.

Caballero and Londono point out that less ambition would not be a solution. Ducking away from the problems would not make them smaller, but bigger. They propagate exactly the opposite: focusing on the whole, overcoming the narrow focus of the MDGs, wanting to reach all people instead of just the (always the same) interested groups and vested development communities. In the circle of stakeholders, this is a battle between traditionalists and innovators. The traditional line is desperate to hold on to classical development aid which seemed ideally expressed by the MDGs. This line of thinking can imagine the time after the first 15 years only as a linear continuation. From this perspective, the sustainability idea appears almost as an attack on money flows, status and decision-making processes, at least as a disruption to be warded off. Caballero's concept is, in fact, at odds with the conventional development industry. Caballero and Londono also propose that the new concept be universal, meaning that it applies equally to the South and the North, and thus tends to replace the division into donors and recipients in favour of a politics of finding solutions together. It is intended to give ecological humanity issues a broader scope than they have ever had before. It is intended to ground new decision-making procedures and processes and to fully engage stakeholders. It should ultimately be concrete and tangible for all people in their daily lives, improving their lives, empowering them, arousing curiosity and inspiring them to participate.

A week before the start of the UN conference in Rio de Janeiro, Brazil, it's coming down to a showdown. On June 13, the leaders' closest aides will arrive there and set about making the final decisions on how to proceed. The SDG advocates call for an open and transparent process. Participation and transparency should be key to a meaningful set of SDGs. Too often already, good missions and engaging concepts have been stranded in diplomatic nowhere or have bureaucratically being cut down. Too often, the diplomatic silos of G77 and China, the US, OPEC and the EU have imposed their institutional interests.

It will be difficult days in Rio. Only a few people have access to the real negotiations and decisions. This is also due to the, in the end, successful but sometimes risky nature of Brazilian top diplomacy. The host of such a mega conference usually ends up being the main person responsible for the failure or success. So the Brazilians thoroughly sound out all the squabbles and positions and draft a final document that they present to the other countries as take it or leave it. In doing so, they succeed. The so-called Rio 2030 Agenda for Sustainable Development, entitled "The Future We Want", consists of 283 sections. Only seven of them (245–251) are the real core. They are the legacy of Rio+20 and thus of the entire Rio process since 1992. In paragraph 248, governments determine that they themselves will take the lead. Not some MDG-like external expert group, not a UN agency, and not governors of blocs are to develop the SDGs, but the governments themselves. This is to build ownership among government and trust among civil society. The Open Working Group agreed between the governments should work transparently and include all interested countries. Above all, it should openly involve stakeholders from business and society. It is called a technical, not a political, working group – a ploy to avoid the usual formal diplomatic power games. This opens up entirely new perspectives. The more than 70 interested states have to distribute themselves among just 30 seats. Germany shares one seat with Switzerland and France. China cooperates with Kazakhstan and Indonesia, Bangladesh with the Republic of Korea and Saudi Arabia. Japan is organising with Iran and Nepal, which was considered the strangest troika of all. Some African countries supported the SDG process from the beginning and were given sole seats. Hungary also had one, representing the EU, which at the time was still internally arguing about the meaning of the SDGs.

Colombia's impulse becomes successful because the proposal and the ensuing negotiation process are completely different from the usual procedure. What succeeds here is precisely what was not originally on the table. It opens a path that had not been visible before and consequently could not be followed. This makes the story of the emergence of the SDGs one of the most important contemporary narratives. The root of success lies in an intervention that resists the internal logic of the existing. It rather is going off-script.

The SDG process has a clear message: (a) Leadership responsibility consists in listening carefully and drawing conclusions with regard to visions, new paths and the overlooked. (b) Opportunities for transformation have their time, windows of opportunity open and close. The intellectual competence for good analysis must be combined with the power to initiate changes with an ethical compass. (c) You have to be able to read policy cycles, and you have to be prepared to take advantage of the opportunities. As a rule, this can best be done off-script, as was the case with the Berlin Wall, Carlowitz and Brundtland, the German nuclear phase-out and Caballero/Londono. (d) Transformations require courageous ideas and the setting of signs beyond one's own borders and the familiar, possibly also beyond cultural imprinting. With Caballero, the sign for the future in the still male-centred world diplomacy is set by a woman, of all people, from a country of the Global South, of all people, from Colombia, itself shaken by war and unrest, of all people, not a professional diplomat, of all people, an outsider among career diplomats, and on top of that, of all people, at a very late and actually impossible time.

3.7 An overdue narrative: the Anthropocene

In the beginning, there was an outrage. When Nobel laureate and chemist Paul Crutzen coins the term Anthropocene a good 20 years ago, he signals a deep sense of inadequacy. He criticises the ponderous rut with which science addresses climate change and world destruction. Crutzen lacks the right sense of urgency for transformation. He has just saved humanity by a razor-thin margin from the catastrophe of ozone depletion and knows what he is talking about.

His term Anthropocene is an angry call to order. Humanity, he says, has long since found itself in a situation where it is as powerful as the geological force of nature itself, but without being prepared, without the necessary knowledge and without the necessary political tools. Crutzen makes his indignation clear with a provocative symbol. He refused the usual routines of his community. He does not write a book, give a lengthy keynote, call for a major conference, hold a special political sessions, or write an appeal. His short text forms a distrustful contrast to the hustle and bustle of political and of the academic life. The proof comes promptly in the reaction of the scientific community. Geology finds itself particularly challenged by Crutzen's interjection because one of its scientific backbone is the classifying of Earth's ages, and certainly, coming from outside Crutzen's interjection feels like a challenge to them. So, the geoscientists commission a working group. This group was tasked to bring the new proposal in line with the established routines of their discipline, in particular to examine how the Anthropocene could be measured in the Earth's rock formations. The examination has been going on for more than 20 years. During that time, the political public has long since made up its mind. The term Anthropocene fits the time we live in. The term is at home in talk shows, seminars and book publications. But, and this is further evidence of Crutzen's outrage, nothing substantial is happening. Those who talk about the Anthropocene show concern without acting. "It's five minutes to twelve", is proclaimed, an almost collapsing state of the world is stated, only to continue the usual, including wolf howling as another popular element of the entertainment industry. As if the Anthropocene was a place of longing exculpation floats along. It is as if ashes are being scattered on the head of humanity. Self-accusation turns into self-affirmation. Instead of being frightened, one simply carries on.

Paul Crutzen meant it differently when he said:

> Unless there is a global catastrophe – a meteorite impact, a world war or a pandemic – mankind will remain a major environmental force for many millennia. A daunting task lies ahead for scientists and engineers to guide society toward environmentally sustainable management during the era of the Anthropocene. This will require appropriate human behavior at all scales, and may well involve internationally accepted, large-scale geo-engineering projects, for instance to "optimise" climate. At this stage, however, we are still largely treading on terra incognita.
>
> (Crutzen 2002)

The phrase "Unless there is" is humble, but expresses a major truth and foresight. It anticipates the 2020/21/ff. experience of Covid-19, also of SARS and MERS before. Basically, Crutzen is saying that human dominance will only go (reasonably) well as long as nature doesn't fight back. This is a profound truth. Humans dominate nature only as long as its rules are played by, but nature has its own, and those can unleash forces many times more powerful than humans.

A volcanic super-eruption is possible at any time. For example, the eruption of the Phlegraean Fields near Naples, Italy, could bring humanity to the brink of extinction. Geologists and paleo-ecosystem scientists are now well aware of what this means for photosynthesis and human survivability. The Phlegraean Fields are

Europe's supervolcano. In recent centuries, the volcano has remained quiet, but signs are mounting that it is rumbling beneath the "Burning Fields". The same is true of the Andreas Fault and Yellowstone caldera in the US or Kamchatka in Russia and many other "dormant" supervolcanoes.

As the depth of human intervention in nature increases, so does the risk. The risk becomes more complex, while human intervention becomes more specific. Interestingly enough, risks do not show impacts that resemble the intervention in their specific type. Most often, those risks are holistic and systemic. Their variables are unknown in advance. The nuclear accidents are only the most prominent example.

By organising what are thought the virtues of modern government – targeted action, operational effectiveness and coherence, justice and control – the prevailing political mindset all over the world eludes the view of the whole. Breaking problems down into small, seemingly manageable packages is considered the royal road to problem solving. But where all the details are known, where all disciplines and sectors run their programmes smoothly, the whole thing so often remains a mystery.

In Crutzen's approach to the Anthropocene's challenges, humans play a problematic role. Crutzen empowers the science and the engineering elites. Only the expertocracy he sees in a position to guide the whole society. Not a word about democracy, participation, the wisdom of the many. From the people – in general – Crutzen's conception demands that they simply follow the guiding elite and adapt to what they are meant to do. His brief intervention, probably deliberately, leaves crucial questions for further thinking. Who decides who is an expert and part of the elite? Are other experts positioned to call in experts? Who would define what is an appropriate means? Who would have to enforce it? Who would take over responsibility in case something goes wrong? What about accountability and liability? Perhaps Crutzen's text is over-questioning here. This does not detract from its historical value in having posed the question of the future of society in dependence on nature in the first place – at a moment of hubris when society believes it can override nature and be equal to the forces of nature. That this is a delusion is obvious. As it is not the first erroneous belief that has moved mankind this delusion is to be taken seriously.

Expertocracy is not, but democratic resilience and advanced sustainability thinking are part of the Anthropocene's narrative (Töpfer 2013 as well as Töpfer, Volkert, Mans 2013). That narrative is due to be told. It is a narrative about a world society that is more dependent on nature than ever before. As this circumstance is ever more less part of the general awareness and attention talking about the Anthropocene is a wakeup call. Digitality increases the availability of data and data granularity, and connectedness of the world. Maybe it also increases the value and impact of sustainability policies. One key for this the way we transform data into information, transform information into knowledge, and knowledge into certitude, leitmotifs and visionary thinking about pathways and options for the future (see Figure 3.1). As another key of similar importance we have to rearrange for Anthropocenic institutions and operations.

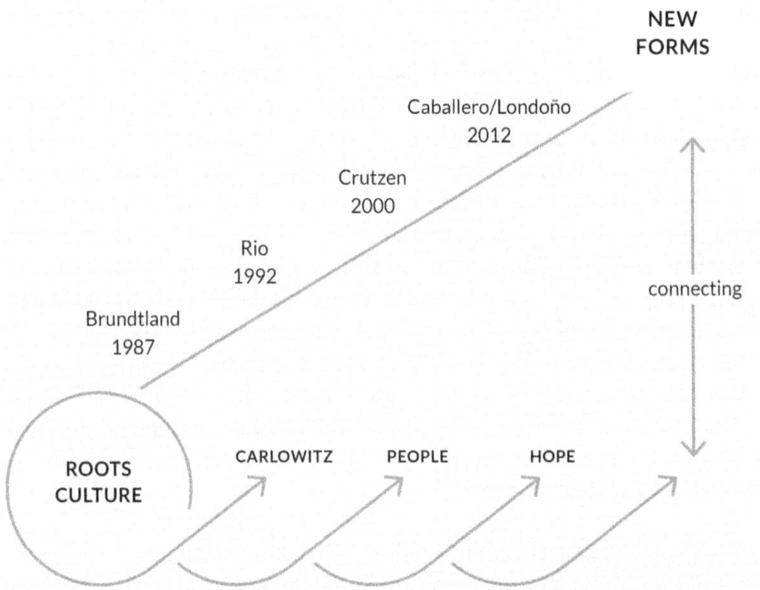

Figure 3.1 From roots to promotors, and to a new collective form of driving forward sustainability

3.8 Growing roots: mindset, governance, knowledge, cultural leitmotif

Transformation is always a matter of power. This is not only true for the cases described in this chapter. It is true in a very general and fundamental sense, as Alfred Crosby (1986, 2003) points out in the history of ecological imperialism and of what in the Americas is known as the Columbian Exchange. In Crosby's historical cases power came in the form of military and – of prime importance – the biological factors, epidemics that affected the course of mankind. In the younger root causes of the recent sustainability movement power comes in the form of public discourse, institutional setup, regulation and empowerment of people, and from reflecting about who pioneered the transformation.

The future works with origin. The conception of futures is based on the understanding of where one comes from. This is shown by the example of the SDGs and the UN Agenda 2030. Once a wild idea, they are now normality and create the space for further thinking. Today, almost all countries in the world report on what they are doing nationally on the SDGs. Countries are engaging multistakeholder groups in one way or another. That it is not without barbs and ignorance is not surprising. Nor is it surprising that the reality is that progress is far too slow.

The momentum for transformation to sustainability is strong when essential elements come together: (a) a sensitivity to pressing problems, (b) ideas that link regulation to self-efficacy, (c) scientific analysis and dialogue, (d) a culture of off-script leadership and responsibility. In short, mindset, governance, knowledge, cultural leitmotif.

Never before could a society truly afford to be as antisocial and sick, as fossilised and stressed by lack of time as our societies are today, and never before has mankind the capacity to end hunger and poverty and to combat global warming. Only, societies suppress this more or less successfully. They pretend not to see the die-off side effects of growth for animals and plants, the depletion of resources and stressed humans. Societies think they are forever. They are mistaken, of course, and any look at history shows that, but still they hold on to the idea of eternity. This is the only way, for example, that such idea can even arise that nuclear radiation waste can be safely stored for a million years; the idea that biodiversity can be maintained despite record numbers of animals and plants becoming extinct, and that we will be able to come to terms with higher Earth temperatures. Those ideas stem from this illusion of eternity. Today, the issue of sustainability is making ever greater demands on politicians and stakeholders. The social mindset is also increasingly taking note of this:

- Every single thunderstorm is weather; all weather is climate.
- Every single ailment is fate; taken together, they are a reflection of a harmful way of life.
- Every single extinct species can be part of evolution; at current rate, all extinct species mean doom for mankind.
- Every single young person's future shattered by poverty is tragic; all together they show the ethical failure of politics.

Credibility and truthfulness suffer from the lack of success and that effects are only marginal. In Germany, we must ask for a change of course in climate policy. Other and equally important issues also call for a change of course, such as the fight against white-collar crime. The German real estate market and German business law virtually invite money laundering and corruption. This, too, damages the credibility of responsibility for the future.

Sustainability strategies are intended to magnify the power of individual sector concerns by leveraging interrelationships and creating the big picture. They should turn these elements – mindset, governance, knowledge, cultural leitmotif – into policy. Therefore, stories and images are needed to satisfy the desire for vision, tangibility, power, and self-efficacy. This narrative resists the logic of the existing. Its "off-script" leaves the safety of prevailing opinion and moves on the shaky ground of bold ideas.

All pioneers were once beginners. From the great mass of beginners we know only the lucky few. But all the others are no less important, especially since those who failed in a particularly intelligent way provided the basis for new, better attempts. Germany's first large wind turbines quickly broke down or did not even start to generate power; the beginnings of the solar industry have nothing to do with today's technical standards; educational projects on sustainability also had to grow into their current quality.

Good ideas often start in a small circle of companies, projects or initiatives. Often there is a certain "tick" to them, which is often seen as a deviation from the normal plan and normal processes. Often they start with an irritating deviation from what was previously considered normal. Carl von Carlowitz escaped the narrow dress of his tasks, Paula Caballero contradicted the mainstream, Gro Harlem Brundtland got rid of the neoliberal laissez-faire. The risk of such beginners is the regular know-it-all response from the public. The downplaying of mistakes is an almost pleasurable exercise, which is ultimately intended to conceal the know-it-all's own lack of action. If sustainability concepts serve even a single purpose, it is hopefully to encourage beginners to learn from mistakes and to turn beginners into masters. To do this, it is helpful, if not necessary, to know the big picture of sustainability policy.

References and recommended reading

Assmann, Jan (2011) *Cultural Memory and Early Civilization: Writing, Remembrance and Political Imagination*. Cambridge: Cambridge University Press.

Avery, Gayle and Harald Bergsteiner (2016) *Sufficiency Thinking: Thailand's Gift to an Unsustainable World*. London: Georg Allen & Unwin.

Beck, Ulrich (1992) *Risk Society: Towards a New Modernity*. London: Sage, first published in 1986 as: *Risikogesellschaft. auf dem Weg in eine andere Moderne*. Frankfurt/M: Suhrkamp.

Bertelsmann Stiftung (ed.) (2013) *Winning Strategies for a Sustainable Future*. Reinhard Mohn Prize 2013, Publisher Bertelsmann Foundation, Gütersloh (responsible: Armando García Schmidt/Bertelsmann Foundation), www.bertelsmann-stiftung.de/fileadmin/files/Projekte/31_Nachhaltigkeitsstrategien/Case-Study-Bhutan_Reinhard-Mohn-Prize-2013_20131016.pdf

Brundtland, Gro Harlem (1987) *Our Common Future, from One Earth to One World*, www.un-documents.net/ocf-ov.htm#1.2; https://sustainabledevelopment.un.org/content/documents/5987our-common-future.pdf

Caballero, Paula (no year given) *A Short History of the SDGS*, www.slidelegend.com

Carlowitz, Hans Carl von und Joachim Hamberger (eds.) (2013) *Sylvicultura oeconomica oder Haußwirthliche Nachricht und Naturmäßige Anweisung zur Wilden BaumZucht*. München: Oekom Verlag.

Crosby, Alfred (1986) *Ecological Imperialism: The Biological Expansion of Europe 900–1900*. Cambridge: Cambridge University Press.

Crosby, Alfred (2003) *The Columbian Exchange: Biological and Cultural Consequences of 1492*. Westport: Praeger c/o Greenwood Publishing Group (first edition 1972), available online on google.books.

Crutzen, Paul (2002) Geology of Mankind. *Nature* 415, 23, https://doi.org/10.1038/415023a

Crutzen, Paul J. and Eugene F. Stoermer (2000) The "Anthropocene". *IGBP Global Change Newsletter* 41, S17–S18.

Diamond, Jared (2005) *Collapse: How Societies Choose to Fail or Survive*. New York: Penguin Books.

Dodds, F., D. Donoghue and L. Roesch (2016) *Negotiating the Sustainable Development Goals: A Transformational Agenda for an Insecure World.* London: Routledge.

Grober, Ulrich (2012) *Sustainability – A Cultural History*, translated by Ray Cunningham. Totnes, UK: Green Books.

Grober, Ulrich (2013) *From Freiberg to Rio – Hans Carl von Carlowitz' 'Sylvicultura Oeconomica' and the Career of the Term‚Sustainable', Lecture*, http://carlowitz-gesellschaft.de/media/2020/03/Von-FG-nach-Rio-englisch.pdf

Haber, Wolfgang (2010) *Die unbequemen Wahrheiten der Ökologie: Eine Nachhaltigkeitsperspektive für das 21. Jahrhundert.* München: Oekom Verlag.

Hauff, Michael von, Robin Schulz and Robin Wagner (2018) *Deutschlands Nachhaltigkeitsstrategie*. Konstanz und München: UVK Verlagesgesellschaft.

Hauff, Volker (hg.) (1987) *Unsere gemeinsame Zukunft: Der Brundtland-Bericht der Weltkommission für Umwelt und Entwicklung.* Köln: Greven.

Kamau, Macharia, Pamela Chasek and David O'Connor (2018) *Transforming Multilateral Diplomacy: The Inside Story of the Sustainable Development Goals.* London: Routledge.

Leff, Enrique (2010) *Latin American Environmental Thought: A Heritage of Knowledge for Sustainability.* International Society for Ecological Economics/South American Environmental Philosophy Section, Publicación Ocasional no. 9, https://iseethics.files.wordpress.com/2011/03/saps-no-09.pdf

Osterhammel, Jürgen (2015) *The Transformation of the World: A Global History of the Nineteenth Century*, translated by Patrick Camiller. Princeton: Princeton University Press.

Osterhammel, Jürgen (2009) *Die Verwandlung der Welt: Eine Geschichte des 19. Jahrhunderts. Historische Bibliothek der Gerda Henkel Stiftung.* München: Beck Verlag.

Reitemeier, Arnd, Ansgar Schanbacher und Tanja Susanne Scheer (Hrgs.) (2019) *Nachhaltigkeit in der Geschichte. Argumente, Ressourcen, Zwänge*. Göttingen: Universitätsverlag Göttingen.

Renn, Jürgen (2020) *The Evolution of Knowledge: Rethinking Science for the Anthropocene.* Princeton: Princeton University Press.

Sächsische Hans-Carl-von-Carlowitz-Gesellschaft e. V. zur Förderung der Nachhaltigkeit (German Only, Divers Publications), http://carlowitz-gesellschaft.de/

Schmidt, Falk and Nick Nuttal (eds.) (2014) *Contributions Towards a Sustainable World: In Dialogue with Klaus Töpfer*. München: Oekom Verlag.

Töpfer, Klaus (2013) Nachhaltigkeit im Anthropozän. *Nova Acta Leopoldina NF* 117, no. 398, 31–40, www.leopoldina.org/uploads/tx_leopublication/Probekapitel_NAL398.pdf

Töpfer, Klaus, Dolores Volkert und Ulrich Mans (eds.) (2013) *Verändern durch Wissen. Chancen und Herausforderungen demokratischer Beteiligung von "Stuttgart 21" bis zur Energiewende*. München: Oekom Verlag.

4 Strategising

The state-of-the-art of German sustainability politics

4.1 A latecomer turns into a front-runner: starting points and uphill struggles

Germany, as a whole, is politically committed to promoting sustainable development on all federal levels. Parliament is active. Research and development programmes are underway. With respect to civil society, academia and the private sector, there are campaigns and structured dialogue processes. Subnational and international networks are actively promoted. The German Government also helps implementing the 2030 Agenda worldwide. These operations are only made possible by setting up what in Germany is called an architecture. That is, in a nutshell, the image Germany paints of itself. It is completed by the demand of the players, be they ministers, heads of associations or campaigners, that "we" must become even faster.

If all the self-image should be correct, then the question remains why there is still such a large distance to the goals, why one is not already further? Why does the demand for "faster" implementation remain so pale? Why do the numerous appeals and calls not take effect?

There is no single reason for this. Rather, it is a bundle of inconsistencies in analysis and deficiencies in implementation. These mismatches are not so severe that they would derail the entire process, but they effectively slow the process down and cause it to meander in a slow flow where a strong flow would be possible.

Before 1998, all attempts for a national sustainability policy failed. As already shown in the chapter on the Brundtland Momentum, sustainability policy did not make decisive progress in Germany in the 1980s and 1990s. It was not until 1998 that politicians agreed on next steps and established a socially relevant gatekeeper and an administrative governance mechanism. Since then, all four federal governments have stuck to the idea of sustainable development. Chancellor Gerhard Schröder until 2005 and Chancellor Angela Merkel 2005–2021 have delivered keynote speeches each year on the status and prospects of their sustainability policies. They gave emphasis to the top-level engagement that is a pillar for any sustainability policy. The annual public conference of the German Council for Sustainable Development has become a yardstick for sustainability policy. In

DOI: 10.4324/9781003242024-4

recent years, the Parliament increasingly debates about fundamental issues of sustainable development policies.

The concept of National Sustainable Development Strategies (NSDS) as first proposed in 1992 aims to integrate economic, social and environmental objectives into one strategical focus, with widest possible participation, a thorough assessment, and ambitious targets and timetables.

After 1992, high level UN meetings repeatedly reflected on the importance of NSDS and called on the responsibility of leaders and all interested parties. However, the outcome kept very poor. The high-end requirements may have been designed in good faith, but with no sense for reality. It is an over-design. In the 1990s, the high-end approach never was never taken seriously. It led to numerous helicopter studies and layouts of ideal processes never to be implemented. This translates into ten to 15 lost years. The most tangible downside was the disappointed of those who honestly believed in the Rio-Agenda and its tools, be it in the position of decision-makers or of ordinary people.

By 2002, only about 85 countries had developed at least some form of national strategy, the nature and effectiveness of which varied considerably. None had substantial success. The devastation of an idea was palpable. Any reiteration of the 1992 Rio demand was perceived a soft funeral rather than a wake-up-call. Policy makers viewed the strategy documents of the time as bureaucratic monsters created by a small group of insiders and not worth much to the real world. This kind of downplaying of sustainability strategies is still around, but not to the extent it once was.

Germany completely missed this first phase. In the 1990s, Germany was so preoccupied with internal issues of reunification that it lost touch with the global discussion on sustainability while European pioneers such as Finland and the Scandinavian countries, but also Great Britain set good examples at this time. Finland's sustainability policy, for example, got off to a successful start right back in 1992. Germany got serious ten years later. As sustainability regained political weight the latecomer became a frontrunner.

Nonetheless, Germany's starting position was no comfort zone. Well into the 2000s, in Germany sustainable development was a well-contested concept. Campaigners for human rights and environmentalists did not even like the wording. To them its comprehensiveness seemed to impinge upon their clear-cut mono-topical campaign motifs. The concept would give the economy at large too much say, environmentalists feared. Economists and the business community added to this misconception by using the concept for greenwashing and "same-same" attitudes. The Green party, at that time junior partner in the red – green government coalition, mostly opposed its inclusive character because they perceived it to be in competition with their green identity. Conservatives denounced the concept as a socialist planned economy and declared its alleged paternalism doomed to failure. The general public was either sceptic or had not heard a word about sustainable development. Good ideas remained in a very tiny niche.

With the change of government in 1998 – a coalition of Social Democrats and Greens took over the government – eminent voices from society, business and politics began to speak out in favour of a gatekeeper concept that would finally push and pull sustainability policy out of the niche.

The Sustainability Council was established as this central gatekeeper, both advising the government and acting in society. But still, there was suspicion that the government is only just covering up its responsibility and that it would bypass parliamentary control. Environmental activists didn't trust the term sustainability and they feared that environmental policy would be set back if the trade unions and some entrepreneurs were also at the table. Today, these reservations are forgotten; back then, they were central. They shaped and showed the self-image of the respective groups more than those groups could have wished. It was all about demarcation and claim battles, not about cooperation and willingness to engage in dialogue.

The development to the current status is not a linear one-way street. Nothing must be taken for granted. It may well go backwards again. Rather, the focus must be on a constant further development of the institutional arrangement. The Germany of 2021, like other countries, is under the impact of droughts, forest fires, severe weather and flood disasters with hundreds of fatalities. Germany also has a number of other homemade problems. At present, they could mean even greater political problems: delays in digitisation, financial debt, uncertainty about Germany's role in Europe and the world, injustice, and a lack of respect for one another.

All too slowly the messages trickle down:

- Every single weather is nature; together they are climate change.
- Each individual extinction of a single insect species may still be evolution, together they are overkill.
- Each individual growth programme and each consumption may be justified, but together and in aggregate they are destroying society's financial future.

Those are the result of a way of life that turns against ourselves. Because this is a complex, ramified, and holistic issue, it requires just such answers. Sustainability strategies are one such attempt. How far have we come?

To answer this question, one must weigh what has been achieved and what is lacking, action and inaction. What looks so simple is – in political terms – a difficult and ambitious task. The parable of the glass that is either half full or half empty is often quoted. Pessimists and optimists are readily distinguished in this way. But the glass parable does not give a coherent picture. It conceals that the glass may be unsuitable as a measuring cup. It is simply the wrong glass. But now, instead of admitting this, both sides insist on using the glass as an objective standard, just to be right in each case and to be confirmed in their position. Both sides, however, are making things too easy for themselves. They turn the distinction between optimism and pessimism into a moral question of 0 and 1, when what is really needed is a sensitive analysis of political trends and opportunities. The following is an attempt to recognise changes for the better without losing sight of the unresolved issues and challenges.

4.2 The German sustainable development strategy

Every society needs to determine for itself, what is its best gateway to sustainable development taking into account the prevailing social context. That is a point for sovereignty. A single "blueprint" for national sustainable development strategies would most probably be running into a black hole, similar to the original 1992 Rio call.

The particular label applied to a national sustainable development strategy is not important, as long as economic, social and environmental objectives are balanced and integrated. Countries, enterprises and cities do not necessarily have to label their chosen instruments as a national strategy. They may use alternative wordings and alternative setting for their strategic considerations. As long as the appropriate ingredients are put in place – targets, timetables, sustainable management, walk-the-talk measures, sector-integrating trajectories, whole-of-government and whole-of-society approach, global responsibility – this is alright.

Since 2001, the German governments use the Council for Sustainable Development as a multistakeholder gatekeeper and as a political pathfinder. Its establishment had been the single most important political step, followed of course by rearranging administrative and other procedures. Almost 20 years of practice have yielded a fair amount of insight into what works and what does not deliver, as well as into the shades of grey that lie between those poles.

The German government regularly updates the German National Sustainable Development Strategy once in a legislative period, which in Germany means every four years. The most recent update of 2021 introduces the concept of so-called major transformation areas that each cover a set of SDGs: (a) human well-being, skills and social justice, (b) energy transition and climate protection, (c) circular economy, (d) sustainable building and mobility, (e) sustainable agricultural and food systems, and (f) a pollution-free environment. The government also added another nine additional indicators to make a total number of 75. The first draft of the strategy that dates back to 2001 counted only some 20 indicators.

The increase in the number of indicators is an interesting phenomenon. It points to a trend of paramount importance that is otherwise unmeasured and rarely mentioned. It can be read as an increase in internal administrative approval and interdepartmental relevance. Nowadays all departments are eager to demonstrate that they have a stake in sustainability and the strategy. They want to show relevance and compliance and therefore produce indicators. This is quite different to first decade of the strategy when some departments avoided participation and ducked efforts to talk them into. Internally, among government officials at the time, the strategy was seen only as an additional workload and basically superfluous. This did not enjoy much sympathy. Among government officials, it was common to make plans to exit the strategy, and to end it altogether. The broad acceptance is a real turn around. It is a multifaceted result of the establishment of commissioners in each of the government's silos, the council's leadership convening power, and the programme of measures for a sustainable administration. Since 2010 the government commits itself to implementing the guiding principle of sustainable

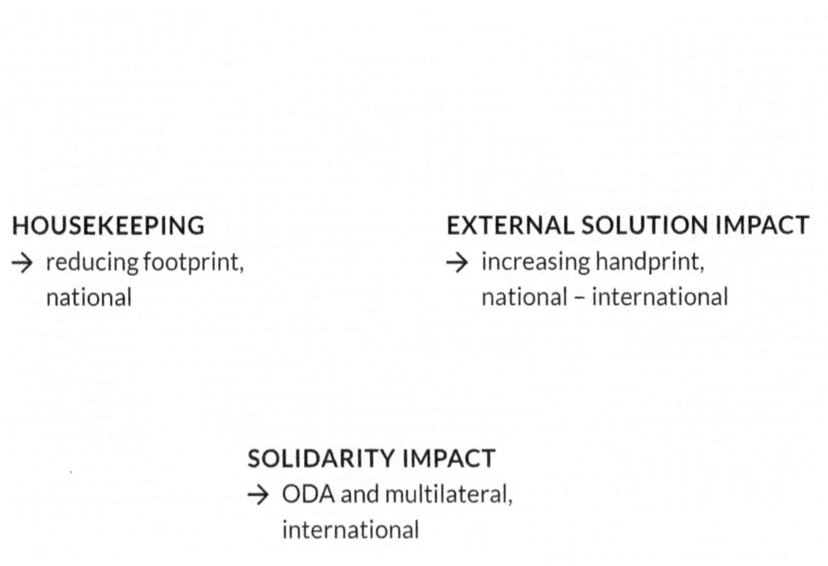

Figure 4.1 Strategic impact mode of the GSDS

development even more strongly in its own administrative actions, fulfilling its responsibility as role model. Indeed, the activities of the public sector has a relevant impact on the dimensions of sustainable development.

Every year, the public sector awards contracts worth hundreds of billions of euros to private companies. The volume of federal procurement is estimated at up to 100 billion euros. The regulations for procurement were only recently revised in September 2021. In the future, greater attention must be paid to energy efficiency and greenhouse gas emissions during the life cycle must be included. A carbon shadow price is to be determined. A no-buy-list applies, for example, to outdoor gas heaters, beverages in disposable packaging and disposable tableware in canteens and at major events. Since public procurement in Germany is organised on a decentralised basis, a new regulation must first improve the data situation. Now the procurement process is also expected to document sustainability criteria.

At first glance, this seems like questionable micromanagement, but it is not. Within administrations, it rather builds the case for a positive self-image of sustainability. Of course, managing a high number of indicators and targets is also a challenge. It raises doubts as to whether a central strategy should cover such large number of policy fields and whether this really helps the political steering process in the chancellor's office. So far, however, there is a mainstream consensus among administrative staff at all levels that the large number of indicators makes sense.

The strategy is supposed to work on three strategic levels (see Figure 4.1). First, there is the internal housekeeping level: with regard to implementation of SDGs throughout the country itself. Second, there is the externality level: with regard to

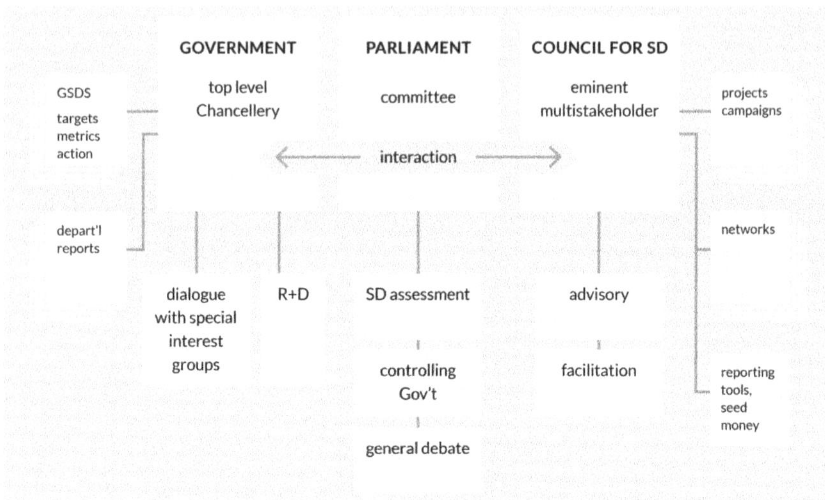

Figure 4.2 Institutional setup

creating technical, entrepreneurial, social solutions in Germany that other countries may co-benefit from (e.g. scaling down specific costs for generating circular economy solutions). Third, there is the solidarity level: with regard to supporting other countries and multilateral arrangements (mostly official development aid and foreign policy).

Since the first resolution of the National Sustainability Strategy in 2002, not only micromanagement, but the governance at large has been steadily developed (see Figure 4.2). The "making of" which has gone through the following stages. In 2000, the red-green government decides to develop a national sustainability strategy, to form a committee of state secretaries, and to convene a Council for Sustainable Development for this purpose and to staff it with relevant stakeholders. For many months, the ministries and departments argue about responsibility for the strategy and who may appoint whom to the council to be created. Several departments claimed a veto power in the cabinet. In the face of this squabbling, Chancellor Gerhard Schröder takes the initiative and assigns the leadership of sustainability policy to his chancellery. In April 2001, the chancellor himself constitutes the Sustainability Council with prominent stakeholders from society, science, business and practice. He asks them to draw up a first set of targets and indicators for a sustainability strategy and otherwise gives the council the right to set its own topics and discuss emerging problems and solutions in society. Chancellor Angela Merkel adopts this governance principle. In 2004, the German Bundestag follows suit and creates a Parliamentary Advisory Commission for Sustainable Development. Its members monitor the German government's sustainability policy.

The Committee of State Secretaries, the Parliamentary Advisory Commission and Sustainability Council form the core of the institutional setup for sustainability policy. In March 2021, the government updated the German Sustainability Strategy. As regards the general elections in September 2021, the Committee of State Secretaries for Sustainable Development ensure continuity in subsequent years. Any new government is advised to deal with the strategy as soon as possible. By the beginning of 2024, it should comprehensively redevelop the strategy with the involvement of stakeholders.

Since 2016, the sustainability strategy has been moving away from the guiding image of the three pillars of sustainability, which makes ecology, economy and social issues appear to be of equal importance. The German government now postulates the existence of "absolute limits" to human demands on the environment, mirroring the discussion about planetary limits.

Most recently, there was a peer review by internationally renowned experts in 2018. They criticise the government's "off track" situations. In response, the government is stepping up implementation and defining further measures and responsibilities. It intensifies the participation of civil society organisations. It also adopts a set of guiding principles for its own work, in line with the previously mentioned walk-the-talk approach.

Assigning the prime minister's office (in the German case the chancellery) the role as centre of sustainability governance is an innovation. It has quite substantive effects, in particular in a country with its decentralised federal structure. Plus, traditionally, the German chancellor's office is not assigned any operational task. This is different from presidential systems, the British system, or the governance of the United States. In Germany, the chancellor sets the great guidelines for government policies, while the individual ministries are responsible for their own activities. All operational issues are the sole responsibility of the ministries. On the other hand, the entire cabinet must reach consensus on all major issues. In governance practise over the recent years in every major political issue it became a habit to call for the Chancellor's office to take over operational control. This was the case in the financial crisis, during the refugee and migration crisis in 2015/16 and in the debate on climate policy although there are quite a few proven instruments for coordination below this level as well.

In the case of sustainability policy, it took almost a decade for the chancellor's office to give itself even minimal operational capacity in terms of staff and budget. Some operational tasks, for example the selection of indicators or international sustainability policy, are still left to the ministries. The head of the chancellery, in the rank of a federal minister, chairs the committee of state secretaries from all ministries. The decisions on the sustainability strategy rest with him. The participation of civil society organisations in the strategy is also orchestrated centrally by the chancellor's office. Another instrument is the formal sustainability assessment. Parliament checks whether draft legislation is in line with the objectives of the sustainability strategy. Technically, it is supported by an electronic procedure provided by the federal government. This improves transparency. However, the personnel and instrumental capacity for those checks is currently still very limited.

Of course, any institutional setup may change. This is also true for the German architecture of institutions devoted to sustainable development and their networks and binding ties in society.

4.3 Gatekeeping institutions

Gatekeeping is a process by which engagement, awareness, and action is kept alive and most desirably even advanced. More technically speaking, a gatekeeper is helping to keep a policy cycle from fading away. This definition differs from the usual sense of gatekeeping which concentrates on controlling access to power. As regards the media, gatekeeping means the process of filtering which news will be published. In our context gatekeeping means to craft actions and bits of information into a limited number of policy messages that are relevant to decision-makers. This is a power function, and choice editing is one of the major gatekeeper features. Ensuring both substantive and reliable options essentially builds credibility and attractiveness.

The executive administration and the discourse-driven societal context follow different characteristics and functions. They tend to display different subcultures that, of course, include organisational principles, concepts and beliefs. For a good and meaningful gatekeeping individuals and organisations need to be aware of both contexts. In the best case they may use the best elements from those two sides.

4.4 Administrative architecture

It is an often-repeated demand that any institutional framework will only produce results if it is building on a substantive and ongoing political will. This, however, is trivial. How could that be otherwise in democratic society? Institutions that work without the highest political backing are almost unthinkable and where they exist they send strong signals of despotic arbitrariness. Political will, this is a lesson drawn from many bad experiences, is not a given, but needs constant support, facilitation and challenge. This is because political will is a double-sided coin. First, it needs political leaders who are interested in sustainability, and, second, it needs gatekeeper that are able to provide the right opportunities for the leaders to exhibit leadership. So, political "will" is always associated with staging the adequate political "opportunity". Both sides of the coin are essential for the administrative mechanism. If one side fails to deliver this shortcoming cannot be made up by the other side.

The German architecture of the administrative framework covers executive and legislative tools as well as data management and advisories.

- Steering element:

  The State Secretaries' Committee for Sustainable Development steers the Sustainable Development Strategy as their key instrument for strategically guiding and implementing sustainability in Germany's policies. In the federal government, the state secretaries have the task of drawing up and preparing the regular cabinet agenda. The cabinet only deals with decision-making issues that have been agreed at the level of the state secretaries. In addition to those regular meetings, and in the capacity of the committee in charge for the SDS, the group runs four to six separate meetings annually, chaired by the head of the chancellery. The departments have to report to the committee on housekeeping in terms of sustainable development. The sessions produce decisions and allow for an open debate on topics for which the committee invites external experts. The council and the parliamentary body take active part as regular guests.

- Control and democracy:

Since 2004, the Parliamentary Body on Sustainable Development controls the Government's policies and politics in this field. It triggers plenary debates and generally raises the level of attention of other parliamentarians for sustainability concerns. It discusses the sustainability assessments (checks) that the government has to produce in the course of drafting legislation, and it can even reject the assessments but that does not stop the legislative process. Even more important is the positive influence that this body has on the general acceptance of the idea of sustainability as part of democracy.

- Using external advice:

The Council for Sustainable Development is an independent multi stakeholder body that gives advice and is given the right and funds to run societal projects out of its own right. The body is set up by the chancellor and reports back to her or him, or to the head of the chancellery. The chancellor selects and mandates members "ad personam", that means by virtue of their professional and personal background, not as a formal representative of organisations. The council acts as gatekeeper although this term is generally not used in its terms of action.

- Data monitoring:

The Federal Statistical Office monitors the indicators of the sustainability strategy in a continued and independent way. It also offers an interpretation of trends and data quality. It gives advice on data coherence and feasibility if the government considers taking new indicators on board.

- ESD, education for sustainable development:

The government establishes the National Multistakeholder Platform on ESD. While the German constitution makes the Länder responsible for the general education (with few examples), the platform works as a supreme steering body. It forms part of the UNESCO ESD-2030 program and has a legacy of successful projects that now have to be scaled up into mainstream structures.

- Administrative arrangements:

  High-level administrators are tasked with promoting cross-departmental joint action. They are called commissioners for sustainability, in addition to their regular task in their respective department. Established in 2017 they are a kind of departmental liaison officers which is a rather new arrangement. Experience has still to be gathered and assessed. The same is true for the so-called principles of sustainability management that are expected to regularly guide administrative routines. Also, as another innovation, the "Science-Policy Interface Sustainability 2030" brings together scientists and civil society practitioners in a first term from 2018–2021. An external evaluation found shortcomings, but clearly encouraged government to overcome inconsistencies and launch a second work period.

- Financial aspects:

  The German government is strengthening the principle of sustainability in subsidy policy and is introducing a specific audit to this end. The German government is also making an attempt to improve the transparency of sustainability targets and the financial budget in general; to this end, it is launching a project in 2021. The sustainability strategy is signalling to the private sector financial market that the issue of sustainability is important by issuing its own Green Bonds and debentures worth 11.5 billion euros. Regulatory measures are being prepared in the EU context. Germany is actively involved in the European development of sustainable finance accounting and disclosure schemes.

- Training:

  Executives and employees of the federal administration are trained with regard to sustainability. To this end, a sustainability education office has been set up in the public administration. By 2030, all federal authorities are to operate in a climate-neutral manner. This stipulation generates a high demand for information and training. A Sustainable Procurement Competence Center provides information on the possibilities of relying on sustainability solutions in public procurement (green procurement). The federal constitution shares political power between federal and Länder level and local authorities, and so, for the respective administrative working levels it makes sense to share insights and manage a steady stream of information.

4.5 Architecture of the social context

A social context for sustainability describes the types of settings in which people are engaged. That includes the groups with whom they do or do not interact. It certainly includes the political culture they live in. Mindsets, customs, routines, attitudes, experiences of falling walls, success and also failure are building up this social context. Like a skeleton they construct the elements of hope and right- and hopeful expectation – on the bright side. The dark side may counterbalance this with despair, disappointed hope, the experience of running against walls, bureaucracy and non- responsive administrations and so on.

In great contrast to the vast literature on how to govern administrations there is absolutely limit writing on the societal governance of sustainability strategies. And what there is fixates on so-called social media and workspace models. The University of Wageningen provides a well-informed website on theories of change dealing with the administrative context, corporate responsibility and multistakeholder engagement. Until today, a parallel practise of change is missing.

The navigation of the social context, however, is necessary for successfully implementing the notion of sustainability in societies as a whole, and with special-interested groups, leaders and activists as well. Intensive and knowledgeable coaching and training is the scarce resource in Germany. The social context of local agenda initiatives, local Rio processes, or entrepreneurial initiatives in that respect was severely weakened in the early 2000s and in many places came to a complete standstill. Protracted negotiation processes, dominating self-interest of civil society groups, neoliberal entrepreneurial thinking, the lack of quick successes and the tenacity of entrenched routines contributed to this. There was little confidence in a new beginning. Financial support failed to materialise. Community management was stuck.

Working the social context is never a smooth sailing. Projects are known to fall through; application requirements are known to complicate things in a way that sometimes is virtually inhibiting any success. Funding channels are cut or disappear. Unpredictable things happen and expose regulatory conflicts and inconsistencies beyond scope. Professional protagonists, however, have much to gain by working out of their comfort zone. They will notice (again, freshly or newly) how small the routine worldview is and how the informed combination of wisdom and humbleness can renew a sense of humanity in practice and values.

Non-executive networks create and maintain binding ties in society. In a society that differs into ever more sub-sociotopes, tribalistic groups and information silos this is of urgent need. It must be noted that the German social context framework does neither work on the basis of large budgets nor is it managed by a professional (cost-expensive) public affairs or marketing mechanism. On the contrary, most of the projects display a low-budget or even no-budget design. The most important elements of the German social governance framework in sustainable policies are the following.

- Community engagement:

  On a voluntary basis and driven only by their personal interest in sustainability options, lord mayors meet on invitation of the council to demonstrate their commitment and consult collegially on what they can do, what political demands to make, and how to address the federal and state governments. This commitment has informally energised local groups. It has laid the groundwork for establishing Regional Networks on Sustainability Strategy (in German: *Regionale Netzstellen Nachhaltigkeitsstrategie*, RENN). Subnationally, they support actors from civil society, business, politics and administration on the regional level who want to contribute to sustainable development in Germany by strengthening and networking their initiatives and competencies.

- Best practise:

Initiatives and projects to achieve sustainability goals thrive on visibility, approval and support. These must be organised in a targeted manner. For successfully launching competitions and calls the calling power of a well-accepted gatekeeper organisation is essential. There are now sustainability awards and best-practice initiatives at all levels. The decisive factor is by no means the amount of prize money or whether there is any at all. What is much more important is a culture of recognition and respect. This can be achieved in a variety of "low budget" ways. The leading prize is the German Sustainability Award (see later).

- Private sector engagement:

In general, engaging the private sector is the most important challenge for building comprehensive networks for sustainable development. Of course, this sector is diverse and the internal differences between companies and industries are large, but even more profound is the difference between the standard model of capitalism and Rhenish capitalism. The former dominates the globe, the latter is practiced in much of continental Western Europe. The models mark very different ways of thinking and acting. In Germany, there is a much more cooperative approach to tariffs, vocational training and government regulations, and corporate social responsibility. Long-term timing sometimes takes precedence over the short-termism of the global capital market. Whether all of this works well (enough) is, of course, an ongoing debate, and navigating sustainability issues is clearly adding an even much greater challenge compared to everything that has been practiced in the past. In 2001, at the same time as the council was established, large companies set up the "econsense" network to help its members embed sustainability in corporate practice. This complements older networks of smaller companies committed to sustainable business and also corporate initiatives promoting climate protection measures. A common denominator among all companies is the need for a robust and meaningful standard for corporate sustainability reporting. It becomes part of the license to operate.

Consequently, coaching and structured support are the most important ways to engage the private sector via non-governmental action. There are two most effective approaches involving the private sector in a structured way, one conducted by the council itself, the other in collaboration with the council. These are the German Sustainability Code and, since 2008, the initiative to award the German Sustainability Award. In the meantime, the demand for coaching and consulting in sustainability issues and skills, mindfulness and foresight is becoming denser and offers become more professional. A market has emerged. Established consulting firms are gradually expanding their competencies and offerings as well.

- Arts and Culture:

Germany's mainstream political culture is not overly inclined toward far-reaching visions for the future. Without these, however, the social context does not function. They must be available in variations and must be able to trigger controversy. Artists and cultural institutions have a special role to play here. There are more and more promising approaches. Theatres, for example, are addressing the issue of climate change. Writing workshops at the deep bottom of empty open-cast mines address conceivable futures. Future workshops put young professionals in the Germany of 2050 and ask how climate neutrality will be achieved. In the movie industry there are first examples of green shooting. TV stations set the pace by introducing sustainability reporting. Scientists are discussing arts and how artists can contribute to research and innovation. Slowly and intrinsically, the concept of sustainability is being introduced to the culture industry, not the least managed by the government in charge for arts and culture and the umbrella organisation of the cultural entities. The pandemic, unfortunately, sets activities back or slows them down a bit as artists typically are precariously employed. However, there is little doubt that the mentioned activities, and others, are bringing the cultural aspect of sustainability to the fore.

- Science:

Sustainability requires the creation of knowledge and the engagement of young people, combined with excellence in research. Therefore, science and the humanities have an important role in the context of sustainability policy. It must open up the mono-disciplinary tunnels as well to inter- and even transdisciplinary approaches. This can best be triggered by basic curiosity and science-policy interfaces. So-called Real Laboratories are increasingly used for solving sustainability issues with social partners, even outside research institutions. As a social subsystem, scientific organisations ensure more sustainability practises in housekeeping processes. Since the early stages of the German sustainability strategy purpose-driven calls for scientific research have successfully expanded the social-ecologic framing of sustainability thinking.

- Public:

Providing forums for public presentation of positions and discourse is an important element of governance. By allowing extensive discussion, they create opportunities for face-to-face conversation, even across diametrically different positions. They also provide an opportunity for recognition of efforts, accomplishments, and sharing of insights. Political leadership finds them indispensable, as do networks and stakeholders that challenge leadership.

4.6 The momentum created by the German Sustainability Awards

Sustainability award competitions are a relatively recent phenomenon in Germany. Most of them emerged after the financial and economic crisis in 2009. An empirical overview is provided by the University of Hohenheim (Gebhardt 2016, 2017, 2020). This analysis currently counts a total of 141 awards on sustainability and/or the environment in Germany. In terms of their type, scope and format, however, they are very different. Prizes are also counted that are limited to regions and/or topics and industries. Not all of them see themselves as so comprehensive as to use the term sustainability themselves. The German Sustainability Award (GSA) is the renowned and most attractive sustainability award (see Figure 4.3).

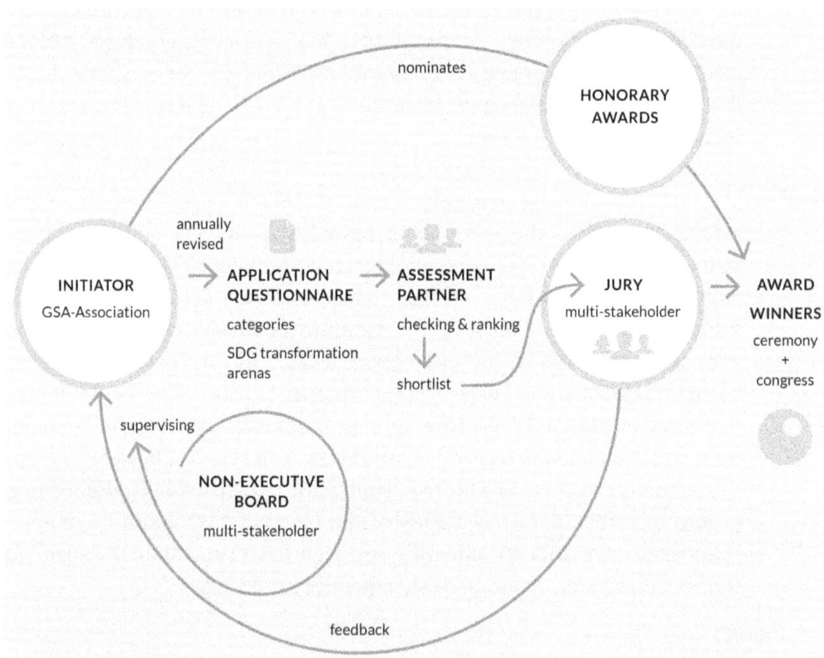

Figure 4.3 Process of German Sustainability Awards (GSA)

The annual awarding has become a benchmark event. This has come about as a result of irritation. For in the onset of the economic and financial crisis in 2008, the idea of establishing an award for sustainable corporate management first appeared as a no-go. Leading industry representatives refused to participate. The German government also did not take the idea seriously enough to take up the role of initiator. The fact that the German Sustainability Award was nevertheless able to be launched in style and with good participation from well-known entrepreneurs is the result of a private initiative that set itself apart from the paralysis that many felt at the time. In Germany, energy policy was not moving forward. Climate protection was micromanaged but not advanced at large. Sustainability was over-conferenced and underachieving. The concept of the German Sustainability Award responded to this challenge and intervened with an approach that was completely unexpected at the time. It embeds a hard competition for best performance in a new cultural form of media event that acknowledges and rehearses the excitement of success. The GSA event is unique. It combines a two-day congress with content and controversy with an evening celebratory awards ceremony. The format thrives on relevant topics. Above all, however, it thrives on appreciation for the best, on the encouragement and motivation provided by the examples and, last but not least, on great emotional moments. Around 2,000 guests attend the congress and awards ceremony. The GSA achieves a high but still to be increased media coverage each year.

Icons of commitment are honoured for their advocacy of sustainability, from H. E. Prince Charles and Annie Lennox to UN Secretary General Ban Ki-moon, then UN High Commissioner on Refugees António Guterres, Gro Harlem Brundtland, Volker Hauff, Klaus Töpfer, UN Administrator Achim Steiner, Lord Mayor Leoluca Orlando, Prime Minister of Bhutan Tshering Tobgay, Jane Fonda, Yusuf Islam (formerly known as Cat Stevens), Livia and Colin Firth, John Schellnhuber, and President Ursula von der Leyen. Celebrities from politics, science and infotainment, who exemplify environmental and social concerns and fight for positive change, give the topic of sustainability and the get-together with the winners the radiance of that big stage that is most appropriate and well earned.

As a multistakeholder project, the GSA involves a total of about 150 civil society organisations and associations since 2008. In addition, partners from science, business consulting, foundations and universities are providing the independent assessment of applications. Changing departments of the German government and the council also support the GSA. Up to eight competitions are held annually, some of which have sub-categories:

Companies (differentiated according to three sizes and transformation fields)
Municipalities (differentiated according to three sizes)
Architecture
Design (for the first time in 2020)
Research
Startups (Next Economy Award/NEA)
Special awards, e.g. for packaging or recycling

The competitions differ in the details of the process, for example if desk top research, an audience vote, a life pitch or interview research is planned. However, it is always about an overall understanding of sustainability and its implementation in concrete contributions to transformation.

The documents submitted by the applicants are subjected to extensive research. From this research, the competition entries nominated for jury consideration are selected. The GSA does not communicate competition submissions that did not make it to the jury's final round, but tries to provide feedback to applicants as requested. For participants the GSA event offers qualified first-hand information and perspectives. It also invites (waiving entrance fees) young people in training or study.

Establishing the GSA initially required overcoming two substantial hurdles. The mere idea of awarding a prize in competition and celebrating the successes of sustainability was met with ideological rejection, with stalling disbelief because the view was widely held that sustainability was a cause for admonition and indictment, but not an occasion for recognition and certainly not for celebration in tuxedo and evening gown. The narrower core of the sustainability community distrusted the social and cultural opening of the topic which was seen as spoiling it in a glittering greenwashed world without depth. Economic and political leaders doubted the legitimacy of awarding the prize and also had reservations about the multistakeholder approach and the overall concept of sustainability.

Ultimately, the success of the GSA belies the initial reserves. The GSA foundation is accompanied by a non-executive board of eminent stakeholders. The foundation, a legal entity as an association, manages the partnerships, financing, event production. It processes feedback and adjustments of the concept, and also proposes the annually composed jury as well as the honorary award winners to the board. The GSA association does not pursue a political agenda of its own, but is guided by the most advanced state of practice at any given time and discursively asks where more can be done and how others can participate.

The GSA makes sustainability a major concern in constituencies and in the democratic self-governance of local governments, thereby broadening the scope of stakeholder action. It helps decisively to save sustainability from the loss of meaning that is feared through inflationary dilution and confusion. It reaches layers and groups of people that have rarely been reached before.

4.7 The role of science and research

The systemic, holistic character of the 2030 Agenda calls for orientation on scientifically based facts and options for action. Specific scientific excellence is just as important as cooperation and exchange among scientists and politics. Still, including practical implementation knowledge in scientific analyses and the transfer of knowledge into practice are challenges, as well as innovative processes, technology and scientific foundations for climate neutrality, emission-free mobility, resource-efficient circular economy, and not least, maintaining social coherence in times of structural change.

In Germany, research for sustainability is a special field of science policy and funding. Over the past 20 years, it has developed from a small community of relatively like-minded researchers into a significant part of Germany's overall research policy. Sustainability science is not a monolithic scientific discipline, but a cluster of different scientific views and approaches.

The German government established this as a special research programme four years after the launch of the political sustainability strategy. Since then updated several times, the programme has funded nearly 10,000 research projects. More than five billion euros have been used for this purpose. Over the next five years, the German government has announced a further four billion euros in research on climate protection and sustainability. The programme builds on the long tradition environmental research has in Germany. Since the 1970s, research has been conducted into end-of-the-pipe filtering and cleaning technologies as well as low-emission production processes. Integrating environmental protection into production modernises industries. In parallel, research into the basic research and applied research into the causes of environmental damage deepen the understanding of the Earth system. Within the concept of sustainability those approaches now receive a broader resonance and an expansion of scientific fields and practises. That is true also for the institutional setup.

The Institute for Advanced Sustainability Studies (IASS) began its work in 2011. Its founding director Klaus Töpfer, formerly Executive Director of the United Nations Environment Programme (UNEP) and a German minister shapes its concept by combining science, democracy and political practice into a debate and new procedures. With Carlo Rubbia, he brought a Nobel laureate to the institute and advanced the understanding of the Anthropocene. Meanwhile, many other universities and research institutions today are realigning themselves and establishing courses of study and institutes on sustainability issues. According to the understanding of the German Sustainability Strategy, promoting research for sustainability will stay an essential part of the transformation.

4.8 On the move: the future of sustainability strategies

Germany, the industrialised countries and the world as a whole are living at the expense of future generations and the environment. This is an equally vague and accurate statement. German Chancellor Angela Merkel, in the summer of 2021 shortly before the end of her term in office reiterates that what has been done to protect the climate is far from enough. This is a depressing truth for her. She insists that the ethical dimension of responsibility must be redeemed. However, such insights still fade all too easily. News of the day washes them away. This is precisely what good governance and sustainability strategies are supposed to prevent, but have so far failed to achieve.

Sustainability and climate politics are cross-cutting issues. Different sectors have to become climate neutral in parallel and with high effectiveness. They normally react to different impulses, and specialist communities working separately from one another deal with them. This increases the need for coordination and mutual control, even in those sectors that usually hardly communicate with each other, or at least not on such issues, such as the construction, agriculture and aviation sectors. The need for coordination is also increasing because of the conflicts of goals between individual measures, such as energy versus nature conservation, urbanism versus car traffic. Even mono-sectoral climate targets are not being achieved, despite the fact that the Climate Act imposes a legal obligation to do so.

The future needs ad institutionalised framework for the cross-sectoral resolution of conflicting goals and the overarching preparation of measures. In Germany, this task is currently fulfilled neither by the sustainability strategy nor by the governance of the Climate Act, the latter after all providing for a climate cabinet, government reports and external expertise, the first with a set of administrative guiding principles, regular high-level meetings on the state secretary level, and formalised reporting and verification of measures. This is not nothing. But it is obvious that the government's routine prefers to issue new regulations rather than address the existing ones that are causing problems.

There is no shortage of proposals from academia and interested parties suggesting even more sophisticated regulation. They focus on control and coordination, and want to build administrative capacity for this micromanaging, with more reporting requirements, administrative burden and controlled interventions. Is this the future, or is a locked-in tunnel view? Based on the global review of national approaches to sustainability, Meuleman (2021) advises a new balance between intra-administrative governance and broader societal governance in working with non-state actors. This proposal goes beyond the usual. The German case proves Meuleman right. Modernisation is necessary because new government services are required. In the past, the state was particularly efficient when it performed separate tasks in separate structures. Now it must combine political and social areas and fields of action that resonate with society.

In the initial phase of sustainability strategies they must be bound to conventional patterns. Usually, this is a command and control governance that works with differentiated pre-scripted procedures and a micromanagement approach. With the initiation phase this can hardly be otherwise. They embed what is new and progressive – goals and quantitative indicators – into a framework of known and well used routines.

When scripted micromanagement doesn't work, the answer is often to step up scripted micromanagement (see Figure 4.4). Governance replies to a problem by strengthening its own logic which is quite understandable. In the end, however, one is locked into a path dependency. In addition to this, there is no evidence that those immanent reactions will lead to the desired result. Rather, it is doubtful that correction is just a matter of being courageous and determined (as it is often called). In businesses, this option would certainly be considered, but one or two other options would also be explored. They would hardly let stranded assets be followed by fresh money.

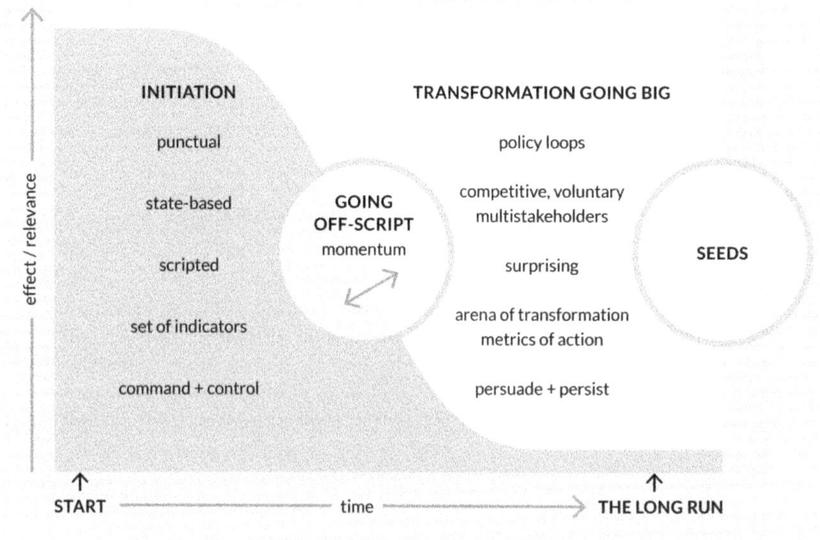

Figure 4.4 Stages of transformation

For transformation at large, the public sector must explore ways and means that would change the way it functions. There are no known historical examples of this. Transformation has always come from the outside. Now the system has to be so radical that it changes itself, from within so to say. Administrations and ministries are not made for going off-script. Therefore, gatekeeper roles must always evolve.

The initiation phase is usefully followed by the phase in which the transformative potential comes to the fore. Feedback loops displace linear processes, advanced (peer review) processes are used for organisational learning. New fields of action emerge, especially in social contexts, for example in sports, culture, communities. Command and control continues to be a useful tool in certain fields, as do quantified goals. Linear command-and-control, however, proves inadequate when it comes to ethical behavior, attitudes, and acting on conviction. The seeds and nuclei described in the last chapter are central to this. However, whether a transition from the initial phase to a transformative phase takes place at all depends on whether a momentum can be created and whether actors succeed (i.e. whether they have the freedom to do so and know how to use it) in doing something new and unexpected.

In a nutshell, the future of both sustainable development strategies and climate politics may be where no-one is looking for: in less scripted details, less prescriptive paternalism, in more basic research, more freedom for new thinking. And, then, roots help us reaching a future where we meet high ambitions and the established goals.

References and recommended reading

B.A.U.M. *Network of Around 700 Companies and Entrepreneurs Devoted to Sustainable Business, Mostly SME*, www.baumev.de/Home.html

BNW. *Federal Association of Green Economy*, www.bnw-bundesverband.de/

Econsense. *Network of International Corporations*, https://econsense.de/about-us/

ESD. *The German Action Plan on Education for Sustainable Development*, www.bne-portal.de/en/german-national-action-plan-1708.html

Fetting, Costanze (2021) *Parliamentary Mechanisms in the Implementatin fo the 2030 Agenda*. ESDN Report, Vienna, www.esdn.eu/fileadmin/ESDN_Reports/ESDN_Report_August_2021_Parliaments_and_SDGs_Final.pdf

Gebhardt, Beate (2016) *Beschreibung von 24 Nachhaltigkeitspreisen in Deutschland mit Relevanz für Unternehmen der Ernährungsbranche*. Institut für Agrarpolitik und Landwirtschaftliche Marktlehre der Uni Hohenheim; Arbeitsbericht No. 26, https://marktlehre.uni-hohenheim.de/fileadmin/einrichtungen/marktlehre/Arbeitsberichte/haa26_net.pdf

German Federal Government (2021) Report on the Implementation of the 2030 Agenda for Sustainable Development. *German Voluntary National Review to the HLPF 2021*, https://sustainabledevelopment.un.org/content/documents/279522021_VNR_Report_Germany.pdf

German Federal Government, website, www.bundesregierung.de/breg-en/issues/sustainability

German Ministry for Education and Research, Research on Sustainability, website, www.fona.de/en/

GSA, German Sustainability Award, www.nachhaltigkeitspreis.de/en/dnp/award/

Hauff, Volker and Günther Bachmann (eds.) (2006) *Unterm Strich. Erbschaften und Erblasten für das Deutschland von morgen. Eine Generationenbilanz*. München: Oekom Verlag.

Lord Mayors Dialogue on Sustainability, www.nachhaltigkeitsrat.de/en/projects/sustainable-city-dialogue/

Meuleman, Louis (2021) Public Administration and Governance for the SDGs: Navigating Between Change and Stability. *Sustainability* 13, no. 11, 5914.

RENN, Regional Hubs Sustainability Strategy, www.renn-netzwerk.de/en/about-us

UNDESA, https://sustainabledevelopment.un.org/topics/nationalsustainabledevelopment strategies

Wageningen University and Research (web) *Managing for Sustainable Development Impact, Integrated Approach to Planning, Monitoring and Evaluation*, www.managing forimpact.org/

5 Keystone

The German energy transition

5.1 The keystone of all sustainable development

The wealth of nations has always depended on the nation's ability to burn things. We have become accustomed to thinking that the amount of energy is essential to the production of social wealth, individual well-being, and political power. Since the beginning of the industrial revolution burning things has become our second nature. The coupling between rates of economic growth and energy demand has produced steep curves. To many leaders, the curve coupling has unfortunately become an axiomatic no-brainer.

Today, however, this worldview has been severely shaken, and rightly so. Profound trends are flattening the curve, and there is reason to believe that the axiom is losing meaning, if it is not already broken. And this is not because the nations lose their hunger for energy. On the contrary, if the nations develop and if politics succeeds in ending hunger and inequality, the demand for energy will increase. Fast growing economies, on the other hand, forces the energy intensity to decline. The continuing shift from industrial to service economies and the increase in energy efficiency comes both with technological innovation and changes in social behavior and consumption patterns. In combination with the growing use of renewables electrification facilitates more efficient technical applications.

The landscapes of fossil and nuclear burning are poised for foundational change between now and the rest of the 21st century. Burning things stop being a sign for progress and, instead, will become an axiom for poverty and being left behind. The wealth of nations will depend mainly on the conversion of the natural forces of the sun (wind, tides and biomass are a result of solar input) and geothermal energy. But we are certainly not there yet.

As a good signal, in industrialised countries the industrial energy intensity is decreasing, renewable energies are gaining and the scaling-up of the renewables market produces a decrease in prices and growth advantages, and on the way is new machinery and digitality with increased efficiencies. The transformation has far reaching consequences for societies, companies and their leaders. The point for both society leaders and businesses is how to build the strategic and political resilience needed to navigate this transformation. Resilience is required because

DOI: 10.4324/9781003242024-5

the trajectory of phasing out fossil energy might be clear in terms of goals, but it is certainly not in terms of practical steps.

Energy remains of central importance to any society. There is no doubt that energy is a keystone of sustainability policies. While there is still unrestricted agreement on this, the consequences drawn from the keystone issue of energy are completely different. Germany and France are going in opposite directions; here they are phasing out of nuclear energy, there they want the phase in into a new nuclear technology. This issue also divides the member states of the European Union into two roughly equal camps. As far as coal is concerned, the situation is no better. Of the industrialised countries, only smaller ones such as Belgium, Sweden and Austria have so far completely abandoned coal. France does not want to build any more coal-fired power plants because of the expansion of nuclear power and relies on the purchase of foreign electricity whenever hot summers reduce the cooling capacity of the French rivers for the nuclear plants and these have to be throttled back. China, India and Indonesia are building hundreds of new coal-fired power plants. Asia has the highest carbon dioxide emissions. Globally, coal contributes around one-fifth of the greenhouse gas emissions. Climate-conscious industrialised countries and development banks in Asia have been pushing for a faster phase-out of coal for years. But so far there is no talk of a general global phase-out of coal. This is even more true for nuclear energy.

The energy transition in Germany is moving forward, although there are serious shortcomings: The expansion of the power grids is faltering, the expansion of renewable energies is delayed and regulated in too complicated a manner, storage capacities and e-mobility are being expanded too slowly in order to make a difference. Energy efficiency is not up to par with what would be technically and socially possible.

The background is as clear as it is dramatic.

The concentration of carbon dioxide in the Earth's atmosphere is 415 ppm (as of 2020). Global carbon dioxide emissions grew by 67 percent from 1990 to 2018. There is no end in sight to this growth. In 2019, about 36.4 billion metric tons of carbon dioxide were emitted worldwide. In addition, there are significant methane emissions. Historically, the old, Western industrialised nations have been the major contributors to global warming. Today, China alone accounts for about 28 percent, and that poses a challenge not only for China (and globally), but for Germany's supply chains and business strategies as well.

5.2 Facts on German climate policies

Germany emitted around 813 million metric tons of CO2 in 2020, slightly less than 10 metric tons per inhabitant. With its one percent of the world's population, Germany thus emits two percent of global emissions. 280 million metric tons of this came from the energy sector, including the production of fuels for transport, 186 million metric tons from industry, 118 million metric tons from heating and hot water for buildings, 150 million metric tons from transport and 70 million metric tons from agriculture. In 2020, 489 terawatt hours (TWh) of electricity were generated in Germany, of which 242 TWh were from fossil fuels.

In 2020, a total capacity of 54,000 megawatts installed Germany generated 51.5 TWh of solar power. Wind will generate a total of 142 TWh in 2020, from 29,608 onshore wind turbines with a total capacity of 55,000 megawatts and 1,500 wind turbines offshore in the North Sea and Baltic Sea with a total capacity of 7,770 megawatts. They supply 112 TWh onshore and 30 TWh offshore. With other renewable sources the total amount of renewable power generated in 2020 is about 251 TWh. Together with efforts in decoupling and efficiency this produces a significant and continuous decrease in greenhouse gas emissions. In 2020, greenhouse gas emissions dropped by 40.8 percent compared to 1990, and by 8.7 percent compared to 2019. Emissions seem to increase again in 2021 due to after-lockdown effects.

The energy sector achieved its sector specific emissions targets, and is actually even an over-achiever. This is mostly due to the decline of coal-fired power generation and the expansion of renewable energy. Coal-fired electricity generation halved only 135 TWh between 2015 and 2020. The share of electricity consumption covered by renewable energy reached around 45 percent in 2020.

The industrial sector, too, achieved its emissions reduction target for 2020 due to the emissions trading and technologic improvements in efficiency. Experts expect further emissions reductions to come from targeted incentives as well as technological developments such as the use of hydrogen in industrial processes. The mobility and the building sectors both miss their 2020 targets. An emissions trading system for those sectors is not yet implemented.

Germany is committed to phase out both the production of electricity by coal and nuclear fired energy plants. The coal phase-out was subject to deliberations of a multistakeholder commission in 2019 involving politicians from the federal level and the Länder as well as environmentalists and representatives of corporate interests, experts, and affected local communities. The commission found a consensus and suggested a phase out until 2038, with a full financial compensation of the affected industries and regions. Building on this consensus, the Federal Parliament and the 16 Länder (the Second Chamber, Bundesrat) adopted the phase-out of coal power and also passed a piece of legislation that provides financial support for the structural transformation in mining regions. The phase-out timeline is gradual and linear. The remaining power capacity provided by coal-fired plants falls to zero by 2038. Hard coal capacities will be reduced to almost zero by 2034. Competitiveness of markets and the carbon pricing might accelerate the shutdown of lignite powered plants. An earlier closure date of 2030–2035 is in discussion.

The true magnitude of the coal phase-out, politically and in terms of materiality, becomes clear only when one considers the installed capacity of power plants in Germany in 2020 (i.e. the plants built, regardless of their degree of actual utilisation, and including buffer and reserve capacities): of 220 gigawatts, about 90 gigawatts are fossil, according to data provided by the German Federal Environment Agency. The coal phase-out will reduce this capacity by 45 gigawatts (the share of lignite and hard coal) by 2038 at the latest, and probably by 2030. This capacity will have to be replaced (and more will have to be added in view of growing electricity consumption). If 2030 is the dead line, decisions have to be taken basically immediately. Currently, it takes about seven years for a wind turbine to be approved and built in Germany, which is clearly not in line with the dynamics of the energy transition. Every previous government has wanted to shorten this period; none has succeeded so far. These figures illustrate the enormous pressure on the industry and politicians. It is a new dimension.

In 2020, civil society activism sets a new standard for ambition. They challenged the 2019 German Climate legislation before the Federal Constitutional Court. The 2019 target path to achieve net zero emissions by 2050 was deemed too weak and the court rejected it in his finding. In April 2021, the court obliged the government to review the phase-out trajectory. It issued a ruling declaring that the Act of 2019 lacked clear provisions for emission reductions in the period from 2031. Specifically, it found the target path would postpone measures that can already be implemented today. It would disproportionately restrict the fundamental freedom of today's younger people. In reply to this, the government agreed on a more ambitious Climate Action Act that is to guarantee climate neutrality as early as 2045 and aims to reduce the financial burden on consumers. The reduction target for 2030 is being raised to at least 65 percent, the reduction target for 2040 to at least 88 percent. This, of course, reduces the sectoral carbon budgets for energy, transport, industry, buildings and agriculture.

In 2022, a new government, formed by a newly formed coalition of the social-democratic, the green and the liberal party, is ready to build on these goals, set additional ones, and implement them with significantly more resources (see Chapter 2.4).

5.3 The European Green Deal is an accelerator

The German Climate Legislation is in line with the European Green Deal. After a discussion phase of 18 month the European Commission launched its proposal for implementing the Green Deal on July 14, 2021.

As of 2035, no new cars with internal combustion engines are to be sold in Europe. Aviation fuel is to be taxed. The number of emission rights used by industry and the energy sector is to be drastically reduced. New targets for energy efficiency, especially for heat pumps, are planned. The commission wants to introduce specific emission schemes for the building and the transport sectors, adding to the EU ETS. It is not to be linked to the existing trading system for industry because the transformation costs are very different. This is in line with the German plan. Finally, a kind of carbon border tariff is being considered for foreign trade.

In addition to taxes, the commission's toolbox includes market-based instruments and regulatory requirements (standards). Overlaps are willed. They serve, according to the Commission, the political resilience of the entire system. The high level of detail in the proposals makes them vulnerable to deviation, but they are the result of the political practice of negotiation, which asks for ever more "certainty" and precision. The scope on fundamental changes is small, as the overall balance must always end up with the 55 percent CO_2 savings agreed for 2030. The final vote for the fit-for-55 package with the European Parliament and the governments of the member states is underway.

As in the case of the German energy transition, the transformation of the European energy system leads into uncharted waters. Basically, this transformation is like a reformatting of the energy hard drive. Different to a computer, the hard drive cannot be taken out and simply replaced by a new one. Here, the reformatting must be done while the system is working. From other differences to the computer, however, the energy system change has a fundamental advantage. This is time and decentralisation. The energy change is not the subject of a one-off, pinpoint action. Experiences, even with problems, can be transformed into innovations. Finally, the energy transition is a matter of many decentralised decisions. Compared to old-fashioned monolithic power plants, decentralised renewable energies empower small local holders and are well-positioned to advance democracy as a whole. What is needed to realise these benefits is to set a direction and to give a framework. This is the purpose for the phase-out of nuclear by 2022 and coal by 2038.

Every day in Germany, someone argues about details and takes legal action against the construction of transmission lines, the planning of wind turbines or solar plants. Lobbies argue about the level of state subsidies that goes to the renewables industry and keep other industries from paying the surcharge for consuming energy. Indeed, the renewables infrastructure is off-target. This often gives the impression that the energy transition is failing at all. In fact, however, this is not the case. Germany is generating more and more electricity from renewable sources, greenhouse gas emissions are falling, new technology is being promoted. And companies want to become climate neutral. To prove this, 69 market-leading major companies, led by the long-established two-degrees foundation, appealed to policymakers in the fall of 2021, clearly calling for more ambitious climate regulation.

A lot of good things could be said about this; only one thing, however, is not: It's not enough yet. Quite the opposite. Heat records and extreme weather situations are becoming more frequent. In July 2021, Germany experienced storm damage with hundreds of deaths that no one thought possible and for which no one was prepared. Drought and heat are killing parts of the forest in Central Europe. Melting Arctic ice is weakening the Gulf Stream's heat pump. The Amazon deforestation is scraping the threshold where self-deforestation takes on and feedback loop dries up the entire country. Climate change deepens poverty and divides the world.

In Germany, attitudes toward climate protection are changing. The number of people in favour of further and tougher measures is rising. It is becoming increasingly clear that the fight against fossil-fuel power plants was still the easy part of climate protection – compared with the task of adapting the entire stock of buildings to be climate-friendly and building up a food system that is climate-positive. People like to discuss coal and cars vehemently, but the hottest issues have changed, with no simple solutions on the table.

The European Green Deal is therefore on the cutting edge. Never before has the Commission announced such an ambitious project and never before has it had the competence and the necessary budget to do so. The Green Deal is broadly based and will give transformative impulses to the economic sectors, but also to the financial industry (sustainable finance) and regulate. The Green Deal brings Europe back into the circle of climate policy to be taken seriously.

5.4 Denuclearisation, a case of German transformative policies

The world is familiar with many types of extremely large and ambitious megaprojects. They include, for example, the construction of the largest dams and waterways, industrial and space projects, or the clean-up of damaged reactors. They are characterised by costs in the billions, centralised individual construction and overarching, systemic effects. The German energy turnaround is such a megaproject, but one without the focus being on a single technical building, but rather cross-cuttingly focussing the whole of society. It is not centralised, but decentralised. It is not state financed alone, but a mixture of public and private finance. The energy transition is Germany's largest and most important modernisation project for sustainable development (see Figure 5.1). To understand the political mechanics it is important to look back into how the energy transition came into being.

From the 1980s until exactly 2011, energy policy in Germany was a matter for board meetings and small groups of experts. Since 2011, it is a matter for multistakeholder processes, small and medium sized companies, cities and local administrations various other drivers. It is people's business. This signals a power shift. Power no longer lies exclusively with a few central energy supply monopolies, but is decentralised and diverse.

The background to this lies in Fukushima, Japan. There, on March 11, 2011, three nuclear reactors burst, shattering a dream in Germany: the dream that unlimited nuclear power would ensure everlasting growth. In Germany, not unlike in other countries, nuclear energy was considered the lifeblood of progress. Nuclear energy was a reason of state. By end of 2022, however, Germany will take the last three nuclear reactors off the grid. The iconic turning point was Fukushima.

→ climate neutrality: 2045

→ phasing out of nuclear: 2022

→ phasing out of coal: 2038

Figure 5.1 Germany's targets

The Fukushima super accident violated more than just a nuclear plant (as severe as this is). In Germany it violates very basic political thinking about progress. This is much more than just another technical disaster after Sellafield in 1957, Three Mile Island in 1979 and Chernobyl in 1986. The live image of an exploding reactor shakes the entire political landscape, including the top of government. The violation of this idea was compounded when it became clear that the reactor meltdown would be exacerbated by the loss of political and technical control: One cools the melting reactor blocks with water virtually from watering cans. Workers try to bridge the broken power lines with car batteries, while cables to the power station are too short. Only by pure chance does a broken flood tank prevent the catastrophe from escalating; no one knows about its leak at first, but its water leaking into the reactor obviously prevented the fuel rods from burning openly, which might have literally torn the Japanese island apart.

Progress is never made of technology alone (although that is often thought). Progress is a web of trust, vision, rules, competence and feedback. Fukushima, on the other hand, delivered lies and masked ignorance, irresponsibility and panic. This is the real message of the accident. Not the technology issue, nor the emergency mode, but the dysfunctional and uncontrolled implications of high-secure technology that goes bonk.

Against this background, the Germany decision is about the ethics of social and political risk management and not so much about the mere replacement of one technology by another.

Critics of the German nuclear phase-out believe that German Angst overreacted emotionally in 2011 and that the rapid nuclear phase-out was unnecessary. Moreover, they argue, it is unwise because the climate problem is more serious than the nuclear risk and nuclear power is, after all, green power. In doing so, they disregard the political background. This may apply, with variations, to all societies – in Germany it is particularly pronounced and plays a major role in politics.

A few days after the reactor burst, the Chancellor announced a three-month nuclear moratorium for the seven oldest German nuclear power plants. She leaves it open how things should proceed after the moratorium. Merkel is following the trend, as many other heads of government are doing the same. They impose similar, time-limited moratoriums and have the safety technology of their reactors checked and upgraded here and there, then bring them back online or even build new reactors. That was also an option in Germany at the time. After all, German reactors generated almost 25 percent of the country's electricity. Shortly before Fukushima, the German government decided to keep Germany's nuclear power plants on the grid even until almost 2040, thus cancelling an earlier government decision. This decision from 2000 contained flexible time limits that allowed the industry to transfer remaining time from one reactor to another. This consensus the then-red/green government had negotiated in close circles with industry. Stakeholders and the then-parliamentary opposition were not involved. When this opposition came into government a couple of years later, one of the first to-do's was to overturn this decision in order to continue nuclear power generation. This happened only half a year before Fukushima.

So, the Chancellor Angela Merkel, on an ad hoc basis, seeks advise by a multi-stakeholder group. Once again, the Chancellor is seeking solutions to complex issues not (not solely) in the expertise of specialised experts, but in broad and responsibly conducted dialogs. She also makes this clear by the name she gives the body: Ethics Commission for a Secure Energy Supply. The Chancellor entrusted the Anthropocene thought leader Klaus Töpfer and the technology professor Matthias Kleiner, then president of a major German scientific organisation, the German Research Society, with the chairmanship of the 17-member body. The commission includes both proponents of the nuclear option and exposed opponents. The author of this book was given the task of advancing the commission's debates through a "living document" that continuously developed into the group's report to Angela Merkel.

The commission advocates a nuclear phase-out by 2022 and proposes a governance framework for this, including a whole-of-society approach, a mechanism to deal with conflicts of legitimate targets, and a dialogue formats that allow for participation, debate and local leadership. The technical shutdown is to take place according to a fixed calendar date (2022). The phasing-out is to complement with a phasing-in of new market designs and renewable technologies that would guarantee that none of the ambitious climate goals would be compromised by the nuclear phase-out. With this, the commission establishes a societal consensus. After some debates government and Parliament follow the technical part of this recommendation. However, they did not implement the recommendations on governance.

As is so often the case: New ideas and concepts will not live up to expectations if the governance is still the old one. Technological modernisation will not work with governance restauration. Fatal inadequacies in the promotion of wind and solar energy as well as the industrial diesel gate and a false start in e-mobility squander trust. Particular interests are emboldened. Organised civil society and academia have accepted this political abdication and, to that extent, are also part of the problem. As a result, the energy transition is working, but it is doing less than it could.

In the crunch time of the non-delivery valley through, doing less than possible is a dangerous thing. Success is never a matter of doing less. Even good ideas can fail completely. To point out what this means, the Desertec project is a good example.

5.5 The nuclear "perhaps" and other mirages

The world is desperate for a technological breakthrough or two that will put an end to all the vexing problems of unsustainability and the hapless attempts to solve them. Of course, it would be nice if there were such breakthroughs: the one big techno fix, and no problems left. Since time immemorial, people raved of the one silver bullet that would make their lives easier in one fell swoop, as promised e.g. by nuclear energy, desert solar power, or hydrogen. At the moment, however, it doesn't look like it. Rather, it is often old or incomplete concepts that are reimagined as mirages. But a closer look makes sense, ultimately also to learn for a sustainable development strategy.

In the late 2000s, Desertec was supposed to deliver the one breakthrough in climate policy. Desertec promised to generate and supply solar power from the Sahara. First, all started with a good idea. Desertec wanted to generate solar power where it makes far more physical sense than in Central Europe. The intention was to transport solar power from the Sahara to Europe via the Mediterranean without any losses. In 2009, politicians, scientists and economists founded the Desertec Foundation. In parallel, German corporations set up the Desertec Industrial Initiative. A brief heyday with many plans was followed by a rapid fall. As a result, companies like Saudi Acwa Power now lead the technology and are very successfully building ever larger and cheaper solar plants. After a break of many years, German industrial companies are now getting more involved again in the Desertec family. This is being triggered by the hydrogen economy. The German government and companies are investing heavily in research and development of solutions here, making them sought-after partners. These are, in particular, suppliers of electrolysers (the technical apparatus for producing hydrogen) and buyers of hydrogen for production of chemical or steel commodities.

In hindsight, the Desertec project turns out to be one of the old concepts of European supremacy. In addition, it underestimated the factors of sustainability. The concept was "top down" and no sense of regional and decentralised needs and markets, politics and stakeholders was developed. There was also mistrust and confusion between scientists, engineers, NGOs, and industry. But surprisingly, the decisive factor was, of all things, German environmental policy. In Brussels, Germany blocked plans for trading in certified green power from credible sources. This trade was supposed to open the renewables market for solar power generated outside of Europe. It would have given electricity customers in Europe the option to choose to purchase green African electricity themselves. Germany saw its national support structure for (its own, national) renewables at risk. This is, of course, a narrow-minded approach that misses the mark when it comes to contemporary thinking about sustainability. Status quo versus future: The shielding of the national renewables market against innovation from outside prevented the entry into a European-African climate protection. The fact that this entry remains more necessary than ever does not need to be explained here. Ten lost years later, large corporations need carbon-neutral energy like an operating license. Some basic material plants, such as chemistry, steel, copper, aluminium, can hardly save any more carbon emissions, due to technical limits to efficiency. Industry's sustainability strategies rely on a change of primary energy. For them, ten years earlier, climate protection was "an option". Now it is an obligation. If you invest in primary industry, you invest in climate protection, or you lose your money.

With regard to the number of industrial mega projects, respectively open issues of a climate adjusted technologies, as e.g. hydrogen, aviation, nature-based solutions, geothermal, biopolymer wood construction and such, lessons must be learned from the Desertec campaign. In particular, this concerns the nexus between resource depletion and climate, bottleneck factors such as water and arable soil, techno fixes and governance, social and peace policies.

Hydrogen technology does not promise the magic of unlimited energy, as is often claimed for advertising purposes. Currently, German policy makers and the European Commission favour green hydrogen from renewables as the only technological option. 62 major hydrogen projects receive a total of around 8 billion euros from the German public budget. France promotes the production of hydrogen from nuclear energy.

The production of green hydrogen alone does not add up to a major swift that would make a difference in the next few years. The maximum amount of green hydrogen currently targeted is like a drop in the bucket. The whole point of sustainability strategies is to look at this big picture and balance more accurately the interactions of technology, culture, politics, consumption and behavior. Altogether those green hydrogen options are neither large enough because they artificially constrain themselves by omitting alternative "blue" options that would produce hydrogen through pyrolytic processing of methane or from natural gas. The methane cracking is a future technology as it allows industry to find new ways to recycle elementary carbon in close loops.

Another aspect of magical thinking is the idea that rapid and disruptive decarbonisation will immediately stop the effects of global warming. The response time of the Earth's ecosystem will likely take a longer time to readjust. It has a long braking period, so to speak.

One of the most significant carryovers from the past is what interested parties call the renaissance of nuclear energy. Fast and dramatic climate change is giving nuclear technology a push, some of the climate leaders pretend. They raise the nuclear Perhaps: Perhaps climate policy needs the nuclear option after all? Perhaps a new generation of reactors are inherent safe? Perhaps smaller reactors can be more flexibly used and perhaps they reduce risks?

What speaks for this Perhaps at first glance is the fact that hardly any country is following Germany's hard nuclear phase-out after 2011. New technology – small modular reactors (where small means 300 megawatt which corresponds with a full-sized fossil power plant) – is said to keep reactors from melting down in the event of an accident and to be less expensive. Theoretic physics might have a point, here. But still its proposed solution is theoretical and needs further technical development and scaling, aside from the embedded "grey" energy that would be needed, while the climate clock is ticking away and emissions are taking precious time off the clock.

France announces to give this nuclear technology a significant boost and is about to develop this technology and to build new reactors as part of their domestic industrial policy. The United Kingdom is betting on conventional nuclear power and is building several power plants with French and Chinese help. Some of the largest nuclear construction projects, however, have outgrown all timetables and financial plans. Likewise projects in Finland and South Africa, for example, are lagging behind their plans and cost calculations. China says it is building 20 nuclear power plants, in addition to all a number of coal-fired plants. Remarkably, there is no private money involved here. Private equity requires fresh ideas and calculable investment risks. An open checkbook situation with a perpetuated timetable is not one of them.

Any even halfway honest cost calculation is to the disadvantage of nuclear energy, and today there is nothing to suggest that this will change in the future. And that's just the cost of building and operating the plants. However, the costs of dismantling the nuclear power plants alone exceed any calculation of economic efficiency and "cheap electricity". Again, take the German example: The four major energy companies operating in Germany have set aside around 36 billion euros for the dismantling of their plants, according to legal obligations and negotiations with the government. 110 nuclear facilities have been built in Germany since the 1950s, including 37 commercially operated reactor units at 27 sites. The last unit will be taken off the grid in 2022 while the dismantling of older reactors is already underway. The estimated costs for the dismantling the Greifswald-Lubmin plant, which was decommissioned in 1990, amount to around 6.6 billion euros, and it is more likely to be more than less. No one has yet included dismantling in the accounting for nuclear energy, nor is there any serious idea of how to calculate and allocate the perpetuity costs of safely disposing of nuclear waste. Financially, this is the real mortgage of nuclear power, assuming that there are no additional extra costs for dealing with accidents, health effects and the clean-up of contaminated sites.

No national economy should afford such expenditures, quite apart from ethical and environmental considerations. The only thing that is safe about nuclear power is that we don't know what to do with the radiation waste and the burdens of eternity. Even if there is an inherently safe reactor at some point in the future, there remain the challenges of fuel transport, securing plants, the societal implications of centralised high-security technology, and the problems with proliferation. After all, the pro climate argument does not sting. The nuclear option is out of sync with the timeline of the Paris Agreement that requires big solutions in the next ten to 20 years – exactly the time that nuclear power plants need for permitting, financing and building – plus the time for the overdue technological reinvention of nuclear plants. For combatting global warming in the crunch time of the delivery gap, the nuclear option is no help but a burden because it blocks money, innovation and intelligence.

Thus, the big Perhaps is a chimera. But it's also true: Opinions are diametrically opposed. Some see a groundswell toward renewable, decentralised, networked and low-risk. The others are betting on nuclear-based projects for the future. The EU wants the Green Deal, but is arguing whether nuclear power belongs in the positive taxonomic list of sustainable finance or not.

The lessons from Desertec and the search for "silver bullet" megaprojects are: a) Technology is never just technology, but always consists of culture and politics. The latter is very essential in most cases and requires political conceptual thinking outside of well-worn tracks. b) Electricity demand is not a constant, it is not linear. c) No national policy, be it a subsidy scheme or a standard, is national. They have a global impact and should better realise this and make it an informed part of a sustainable development. d) Sustainability strategies promote an understanding of the interconnectedness of individual aspects and make it possible to identify and resolve conflicts of vested interests. Without them, long-term solutions are hardly conceivable.

5.6 Goal washing

Across party lines, Germany pursues climate policy mainly with a mixture of enthusiasm (goals, technology) and German Angst. In the fall 2021 general election, a large portion of voters express fear of climate change and favours more stringent and government-controlled measures, while another portion rejects measures that would restrict personal freedom while supporting industrial innovation as a means to address carbon emissions. Experts frequently suggest that just providing more climate disaster evidence and more data would make people change their behavior which is a nice assumption that has not yet proven true. These ideas are quite far removed from real politics, and politics is really only seen by them as the area that always fails when it comes to implementing what they feel is needed. Meanwhile, the US, China, Russia and the United Kingdom are making sustainability a geopolitical factor on par with oil and gas. Sustainability is thus increasingly becoming a question of sovereignty in geostrategic system competition. This competition is no longer perceived as a "green" race, but as the race. Both aspects – idealism and hard geopolitics – are not at all mutually exclusive. Both are entitled and must rather learn to get along with each other. What really weakens the respective position is trying to do one without the other which is exactly what too often happens and what causes the delivery gap. The valley of non-delivery also applies to the German energy transition. A country-wide stocktaking of climate and energy policies, as of January 2022, has a sobering result. The "traffic light" coalition is facing a drastic backlog which leads to a failure to meet climate targets for 2021 and also makes non-delivery likely for 2022 and the subsequent year.

The definition of targets is exhausted. Political parties, associations and the scientific community, up to now, favoured a veritable overbidding competition in terms of goals, without much change in reality. They debate tougher and more ambitious targets whenever the previous ones had not been achieved. This is similar to an attitude that we know from our personal life: For the New Year, we resolve to lose five kilos, which then doesn't work out. Then, the next time, you plan to lose eight kilos. This is a special kind of greenwashing. In a way, you could also call it "Goal Washing".

Politically spoken, goal washing is easier than dealing with the ugly conflicts arising from renewing the energy infrastructure. In the same way, journalistic reporting often makes things too easy for itself when they turn every citizens' resistance against wind turbines and power lines into a heroic story, even if it is only an example for the notion of Not-In-My-Backyard, or it only is a token for an anti-elite campaign against "those above" (a German saying "die da oben").

Goal washing is a leitmotiv that can be observed in other cases as well. Experts calculate that the total amount of renewable power that would be needed in 2050. The amount generated in 2020 (251 TWh) would need to be quadrupled – to just decarbonise a couple of industry processes such as chemistry and steel. Decarbonising the total of industrial and commercial activities, leave alone the full society, would then require another multiple of renewable power. Can that be for real, or is it magic thinking? The siting of additional renewables, onshore as well as offshore, is already today in conflict with nature conservation and land use. The extrapolated demand exceeds any scale imaginable today.

The new German "traffic light" coalition sets out to end the goal washing and to deal with the hard measures. It begins its term with the ambition of tripling the pace of emission reductions to meet the 2030 targets. They want to achieve significantly more in precariously less time by taking action that matters in terms of, e.g. solar on roof tops, extra space for onshore wind turbines, controlling the soaring electricity prices by taking the payment for renewables off the individual power bill (and, instead, using the federal budget money), installing renewable heating appliances, doubling green hydrogen power production. While this promises initial progress, fundamental conflicts of vested interests, e.g. between nature conservation and overconsumption, remain unresolved.

A complete and final phasing-out of burning coal and oil will be a huge challenge for leadership and democracy. But, still, we have to realise that these are low-hanging fruits compared to the challenge to free individual consumption from carbon footprints. It seems quite clear that sustainable development won't work without a substantive reduction of the total amount of energy used in consumption and production (even if it is generated from renewable sources). In the past, transforming the energy sector was the exclusive purview of the industrial energy sector. Increasingly this turns now into transforming consumption patterns into a post-fossil carbon chemistry that works in a cycle and does not pollute the atmosphere.

5.7 Rediscovering nature

In earlier times, people often argued about the difference between weather and climate: not every harsh winter and flood should be misinterpreted as an omen of climate change. Climate change still seemed far away. This has turned. Now the weather is making it clear that climate matters and that better adaptation is urgently needed, and more people than ever before do care are interested in this topic. It follows that policymakers must pay more attention to adapting land use, forests, infrastructure, and the entire food and health system to the changes brought about by global warming. The discrepancy between adaptation and mitigation needs to be addressed and mended. Adaptation must not be understood as a defensive approach alone: of course, it has primarily a defensive function. Adaptation measures attempt to repair infrastructures and assets that have been damaged or are at risk of being damaged by wind, water or heat waves. Because of its nature as runaway problem, the world will never fully grapple with climate change by only adapting to its impacts. Politically, adaptation is always the last straw when nature is perceived only as a victim reacting to man-made changes. The conventional view is that mitigation is about technologies, innovation, economics, and consumption of goods; it is considered a hard option. Adaptation, on the other hand, is conceived as dealing with soft and repair factors, and not seldom also as counterproductive to the need of urgently phasing-in alternative energy options. This dichotomy is short-sighted.

On the contrary to this dichotomy, nature must be made the binding element between mitigation and adaptation. In peat lands, tropic and boreal forests, in soils and in the oceans, nature holds stocks of carbon and methane that cannot not be recovered once released through global warming. Conservation and careful land use can not only safeguard nature's function as a sink, but actually expand it. Nature-based solutions for offsetting greenhouse gas emissions, if properly managed, can serve as a defining icon for the common cause of combating climate change and biodiversity loss at the same time.

Nature based solutions are as urgently needed as the expansion of renewable energies. Ultimately, climate protection can only be achieved if it is achieved on global scale, which means that a great deal of money from the Global North must be used to enable the right measures to be taken in the Global South. This trade-off cannot be made on the basis of high-tech technologies alone.

While this statement is trivial, it is not the basis of current policy. There is a fear that a manipulative practice could circumvent climate goals by unreliable and nondurable measures. Business leaders fear that compensation of their carbon emissions would not be reliable in the eyes of third parties and that the public would brandmark them as greenwashing. That are the reasons why in Germany nature-based climate solutions are not even talked about in a serious way. Consequently, the 2021 German climate law gives the impression that full climate neutrality can be achieved with renewables, hydrogen technologies, and energy efficiency, only, and that those measured would be implemented domestically. The law does not even mention offsetting, nor does the sustainability strategy, although there is already a market in full swing. It is called the voluntary, in order to contrast it to the regulated European emissions-trading scheme. Globally, the voluntary markets are opaque, hidden and quite the opposite of a transparent and fair exchange.

German companies and institutions recently advertise statements such as "first climate-neutral food retailer", "we act climate-neutral", climate-neutral plastic garbage bags, "climate-neutral premium heating oil", "climate neutral product". Cities such as the metropolis of Frankfurt decide to get climate neutral by 2035. The federal and state administrations run a "climate-neutral administration" programme. A well-positioned German operation, the Atmosfair company, recently offers investments into its own pilot scale plant generating jet fuel from renewable sources. As for the offers in general, the market is confusing. They cover a range of measures from small scale low technology such as solar ovens to forest planting or educational and social projects. In general, the marketing claims stand for themselves and do not provide a link to the purchase of credits and certificates, and their safeguards, respectively. Nor do they disclose priority measures to save and replace fossil fuels. Not seldom, the location of the carbon saving, the overhead costs of the supplier, or the social implications are more decisive for the price than the actual carbon storage. Court rulings have already demanded compliance with legal transparency requirements.

Those claims demonstrate that, first, there is already a significant demand for compensation, and this is foremost caused by reputational aspects. Second, competition is on public relation claims, not on the materiality of how to achieve climate neutrality, with climate neutrality a non-defined term. Third, a market that allows unreliable and vague transactions is posing a risk itself for serious market entries, and for the biodiversity. Forth, the suppliers do not provide clear and defined portfolios of nature-based solutions to choose from.

For any public or private entity, it makes sense to offset residual emissions, with residual meaning those carbon emissions that cannot be mitigated by best practice technology. The instrument of carbon contracts for difference presents a strategy for decarbonising large industrial sources of carbon emissions while allowing the companies to stay in the market and to keep up with carbon emitting competitors. This new instrument, namely for the German primary steel industry, is expected to serve as an anchor for a renewable hydrogen, but also leaves room for compensation schemes, depending on what is defined a market compatible decarbonisation technology. To do this, transparency must be created to the way in which the carbon sinks are scientifically calculated and what buyers and suppliers are expected to do in order to ensure the nature-based solution will function over the long term. If this is not the case, the competition remains unfair and unproductive as well which is not a good idea while the European Green Deal and Germany's climate measures will cause the regulated market's price for a ton of CO2 to rise significantly and rapidly. Prices around 100 euros and higher are within reach. You will reach the threshold where the pioneers and front-runners among the companies will be "rewarded" and at the same time it will be the case that the majority of companies will realise that a different time has begun. The same is even more true for the broad majority of consumers and for the popular to populist mass media. Companies will want to become carbon neutral even more decisively and quickly. Demand for carbon offsetting will exceed supply. This trend is already visible at present.

This gives much reason for efforts to mature, clarify and scale up the market as it can be assumed that demand will increase and supplier need more reliable signals for putting together significant and guaranteed project portfolios. It takes no wonder that there are different views on how to deal with regulating nature-based solutions echoing some fundamentally different concepts for regulation (see Figure 5.2).

Based on these differences, another political conclusion is obvious that for the first time assigns a role to the state or an independent entity implementing the due diligence function of a gatekeeper. That would of course innovate governance and politics. Also, this would bridge the controversy over voluntary and regulated approaches. The innovation would introduce a new trust mechanism.

	THE MULTI-STAKEHOLDER VIEW	THE NGO VIEW	THE CORPORATE VIEW
SOURCE	WEF and WBCSD (convener) NCS Alliance: guiding principles for investing in NCS	conservation ---international principles for investments in NCS	McKinsey and WEF consultation paper on the role of NCS
RATIO	only high integrity of NCS makes them effective	NCS projects must equitably benefit community and indigenous rights	build trust around a regulatory framework for a growing market
CONDITIONALITY (HIERARCHY)	NCS only in conjunction with the GHG emissions mitigation hierarchy, and only as interim solution for hard to abate emissions but not permanent	NCS eligible only for companies that disclose carbon emissions, are making progress toward climate targets, keep cutting emissions	policy makers must focus on turning national and corporate carbon-reduction targets into actionable plans, underpinned by binding regulation
PROCESS	robust accounting, verification	negotiated agreements with host communities and governments verification thru third parties	centralized carbon exchanges are needed; carbon emitters should agree on high-quality NCS, advance good practice
COMPLIANCE		comply with and support nat'l host governments policies, carbon claims and accounting scheme	comply with a (to be) commonly accepted international-standard on NCS certification
STANDARDS, SAFEGUARDS	environmental and social safeguards for reductions and removals need to get verified (real, rigorous, quantified) additionality, leakage and permanence appropriately addressed	standards must be rigorous additionality	needed: standards, infrastructure and financing, including carbon reference contracts
ISSUES	avoid double counting, avoid leakage	industry might undermine the Paris Agreement	credibilty issues, company carbon claims lack solid foundation market architecture incomplete, project pipeline needed voluntary and compliance carbon markets should be integrated, international transfer of carbon credits should be made possible

Figure 5.2 Different views on nature-based climate solutions

A trusteeship would audit climate services uniformly and purchase them globally. These would then be passed on to interested institutions and companies as a securitised service in return for reimbursement of the costs plus a surcharge for additional climate justice activities. Examples include the renaturation of peat soils, sustainable agroecological farming systems, blue ocean measures, permafrost management, the establishment of forests, and the introduction of closed-loop recycling systems. A climate trust would improve and enable hard cutting of emissions in the Global South.

This governance could be instrumental to implement the Paris Climate Accords in a way that is consistent with the spirit of the sustainability agenda. Nature-based climate solutions are the cutting edge for climate neutrality going global. For the Global North it is mandatory to channel investments and funds into the Global South. The exchange of carbon sinks and financing for development is the one gateway to a common future.

References and recommended reading

Agora Energie (2021) *Diverse Studies and Papers*, www.agora-energiewende.de; www.agora-energiewende.de/en/about-us/agora-industry/

Doda, Baran et al (2021) *Voluntary Offsetting: Credits and Allowances, Study Commissioned by the German Environment Agency, by Baran Doda, Stephanie La Hoz Theuer (adelphi), Martin Cames, Sean Healy, Lambert Schneider (Öko-Institut)*, free download, www.umweltbundesamt.de/sites/default/files/medien/5750/publikationen/2021_01_11_cc_04-2020_voluntary_offsetting_credits_and_allowances_1.pdf

Düren, Michael (2017) *Understanding the Bigger Energy Picture, Desertec and Beyond.* Cham: Springer International Publishing, https://doi.org/10.1007/978-3-319-57966-5

Ethics Commission for a Safe Energy Supply (2011) *Germany's Energy Transition – A Collective Project for the Future*, Berlin, May 30, English report available from the web archive of Federal Chancellery, https://archiv.bundesregierung.de/resource/blob/656922/457334/784356871e5375b8bd74ba2a0bde22bf/2011-05-30-abschlussbericht-ethikkommission-en-data.pdf

German Energy Agency (2021) *Towards Climate Neutrality*. Pilot Study, English Summary and German full text, www.dena.de/en/newsroom/news/dena-pilot-study-towards-climate-neutrality/

German Industry's Foundation 2Degrees (2021) *An Implementation Programme for Climate Neutrality*, www.stiftung2grad.de/en/

German Ministry for Economic Affairs and Energy (2021) *Dossier on Energy Transition*, www.bmwi.de/Redaktion/EN/Dossier/energy-transition.html

Nachhaltigkeitsrat (RNE) and Leopoldina (2021) *Climate Neutrality: Options for Setting the Tight Course and Ambitious Delivery*. Position Paper, Berlin, June, www.nachhaltigkeitsrat.de/wp-content/uploads/2021/06/RNE_Leopoldina_Position_Paper_Climate_Neutrality.pdf

Prognos, Öko-Institut, Wuppertal-Institut (2020) *Towards a Climate-Neutral Germany: Executive Summary Conducted for Agora Energiewende, Agora Verkehrswende and Stiftung Klimaneutralität*, https://static.agora-energiewende.de/fileadmin/Projekte/2020/2020_10_KNDE/A-EW_193_KNDE_Executive-Summary_EN_WEB.pdf

Schmitt, Thomas M. (2018) (Why) Did Desertec Fail? An Interim Analysis of a Large-Scale Renewable Energy Infrastructure Project from a Social Studies of Technology Perspective. *Local Environment* 23, no. 7, 747–776, https://doi.org/10.1080/13549839.2018.1469119

Van Son, Paul und Thomas Isenburg (2019) *Energiewende in der Wüste. Die Vision ist bereits Realität*. München: Oekom Verlag.

6 Value

The undervalued component

6.1 Value

Value in our sense is the soft form that can add resonance, meaning, gravitas, or resilience to hard administrative, regulatory, or technical measures. This has nothing to deal with the economic attempt – the theory of value added – to explain the exchange of goods and services, and how prices play a central role in organising this exchange.

Value can come in the form of images, music, dance, stage performances, narratives, people's stories, episodes and of course all other appropriate works of art. Highlighting the lives, skills and work of others can also add value that cannot be achieved in any other way. In the context of sustainable development, this added value is best described as cultural framing or as a cultural dimension. So not to be misunderstood: No one can or should misuse the cultural dimension of sustainability to reject or relativise shared global responsibility. This would be the case, for example, if genital mutilation were justified as a cultural trait, or if girls were denied access to education and this was allowed with reference to the cultural diversity of population groups.

In the context of sustainable development, on the contrary, adding value focusses on empowerment of the weakest and on people's sense of urgency and general awareness of issues and actions, and it can be even more important as part of the political culture that helps shape and implement policies. This can be explained by the example of conferences and meetings that are the modus vivendi of the global sustainability community.

Conferences and meetings can benefit from this the addition principle in terms of meaning and significance, and even as regards joy and togetherness. Until today, this is rarely seen or experienced. Here, we are at the very beginning of something new, and essential. The usual conference rooms are boring and rarely convey an idea of why and where people meet and what this means to the outside world. It is one of the most memorable and damaging experiences from hundreds, if not thousands, of meeting events that the structural, temporal, design framework never matches the content of the event. It is even worse when this is somehow recognised subconsciously and causes people to make lengthy explanations about the facts and problems, known to all, that justify

DOI: 10.4324/9781003242024-6

the meeting. Since the advent of cell phones, the posture of the participants of such conferences is the same everywhere. Instead of rapt attention and a willingness to take notes, people sit back, relaxed, and look into their cell phones. Now even many people may be capable of multitasking; answering one's own mails, following chats on social media and listening to what is being said in the room is something more and more people are managing to do because it is also just a matter of practice. However, it also causes speakers to feel like they are in a tunnel, alienated from those they are trying to reach, and waiting in vain for immediate tangible feedback from their audience. They feel compelled to present more facts, figures, and data to underscore the urgency of their appeal. But you don't create a sense of urgency for, say, the number of endangered species by appealing to a sense of urgency and retelling the science. It rarely works that way. Numbers are the friends of data, but the enemies of knowledge and wisdom. It's a strange spiral: responding to the boredom factor only makes a conference more boring.

The worst thing, however, is an artificially "spruced up" conference. In this case the routine of talking heads is interrupted by, say, some artistic interludes that have no substantive connection to the topic under discussion. The arbitrarily selected "beautiful" contributions receive polite applause, at best. However, they do not open minds and hearts.

You seldom find outspoken opponents against this concept of culture, but a great many opinions that reject the concept of Cultural Values, see it a bit on the esoteric side, or consider it superfluous and a waste of time and resources. The community is well versed in enumerating the alphabet of implementation of technical goals and economic measures for which we need capacity building, technology transfer, technical support, data support, political will, alignment and coordination, South-South and other forms of cooperation, gender mainstreaming, incorporation of traditional and local knowledge, public procurement, private sector engagement, public awareness and participation, and transparency, to name the most pressing ones. While this list is already long, one of the main considerations brought into by developing countries remains financial resources, and they are right because the world is nowhere near the suggested numbers. This long list reflects very well the common denominator of the global sustainability community. The bad news is that this, even it is manifold and complex, is hardly enough.

In the discussion about goals and means for sustainable development, cultural values and their contribution to hard wired governance are the most underestimated and neglected aspect. Positively speaking, cultural added value is a constructive component of any successful governance idea and practice. The challenge is to actually recognise this value, accept it and defend it against its opponents, and then realise it in an appropriate manner. "Appropriately" is to be emphasised here because it is often not so easy to find artists or activists who are willing to play along. Within their community, there are voices that reject any notion of a subordinate contribution or "added value" of art. They prefer to insist that art is the only real source of value creation.

6.2 Governance and the cultural codes

Governance is boring: No political professional likes to be lectured about governing. They don't say so directly, but let it be seen indirectly through a certain indifference and their staying away from nerdy governance discussions. They suspect those discussions to have a hidden agenda and to actually want other contents and ends, but do not say so, while they "only" discuss instruments and means to the end. Governance is almost always associated with academic thinking, at least with a theoretical thinking that is not tied to the materiality of its causes but is alienated from them.

Governance is an exciting topic, on the contrary, the governance experts do not tire of emphasising. For them, governance is the key to contrasting new forms of political governance with traditional ways of managing politics. The European Union's multilevel system of governance serves as the main field of competence and experimentation, with its open agenda-setting methods and its negotiation procedures that complement soft and "hard" implementation tools such as guidelines, directives, and infringement procedures to ensure compliance with EU law and European institutional procedures.

Both aspects apply and are far from mutually exclusive. They can even play a role at the same time on the same factual object, and they can build up each other's momentum. The tension created by both aspects (both camps of protagonists) can at times seem curious and Kafkaesque, but also a productive source of innovation. A third aspect is added, probably even the most important: Governance issues are of particular interest to advocates of new policies that have not yet found a politically accepted entry and place in administrative competencies. The latter is undoubtedly the case with sustainability.

This leads us to a weighty point. Governance never stands alone, but must always be seen in the context of individual people (and institutions, respectively) and their role in the domestic political culture. Both aspects cannot be separated. An example: What an environment minister declares at a high-level conference somewhere in the world and what he or she later on can enforce in the national negotiation process may differ somewhat. This disparity is not new, of course, and it is repeated often enough. One of the highly unproductive and most liked standards in the discussion on governance is that this disparity is not challenged head-on, but instead space is given to the notion that "we" need an integrated approach and that this will require a restructuring of the global governance system. On first glance, this seems to be quite obvious. But a closer look reveals a cultural mechanism that is part of the stalemate sustainability politics is in.

This mechanism is almost exemplary, we can find it in many varieties and on many occasions. It transfers the message to an impersonal, diffuse "we" that has no explicit recipient. This mechanism makes the sender bigger than the message. It usually receives broad approval because it does not force anyone's hand. The author of the message makes the recipients accomplices in feeling concerned and doing nothing.

Another often heard traits lies in the saying “as xyz alone (a minister, a country, a community, a sector) we cannot change the status quo/safe the world”. It is the equivalent to the first mechanism, but here the sender dwarfs the own responsibility, and, too, makes the audience an accomplice.

A last example: A politician who succeeds in taking into account the interests of all minorities is usually considered a culturally sensitive, good politician. This politician is said to leave no one behind and to balance the cultural diversity in a fruitful and just way. Undoubtedly, this is truly a virtue and a good commitment, if the politician really succeeds in doing that and manages the heterogeneity of interests. But still, it is again the cultural points that may raise doubts: What good is it if the sum of particular interests does not contribute to the common good? And who would tell the difference?

The sum of the common good is more than the enumeration of the individual points of interests in well-being. This is the addition paradox, and it is well known. The subtraction paradox, however, is not as well known and perceived while it is the more striking and important one. It says that the whole can also be less than the sum, meaning the common good us less than the sum of all (vested) and individually accepted interests. It is even very likely if the limitations to growth are being respected and calculated in. No diversity devoted “good politics” can deal with the subtraction paradox. Only those strategies for sustainability can turn this insight into policy that have a cultural core and use it consciously. If anything, it is only here where a strategy gains political weight and momentum.

The culture core is about how to deal with risk and chance, where both of them are not only made up of numbers and probabilities, but are also reflected in the behavior of individuals and groups. As we know it from our own individual life, culture becomes most evident “when no one is watching”, that is, when control and supervision do not per se produce the behavior those procedures themselves “desire”.

The following points may turn cultural codes into concrete change making. A strategy should, building on legitimised target and metrics, promote the function of a trustworthy gatekeeper, be transparent with regard to conflicts of interest, and also open up political space for procedures outside the rulebook, and may use iconic features that encourage people to think outside the box.

6.3 Transformative practise

Thinking doesn’t sadden. It does not demobilise, but modernises. Practical examples from the German context are available: the positive push for local and regional networks of activists, the 30-hectare-goal to limit the consumption of arable land for settlement purposes, the expansion of organic farming, food waste initiatives, resource productivity, and last but not least, the German Sustainability Code (DNK) as a guideline for corporate reporting (see Figure 6.1). The cultural shift is underway, as evidenced by the fact that sustainability is no longer a dirty word in the business world, and leading print media no longer view sustainability as an ephemeral niche that misses the mainstream.

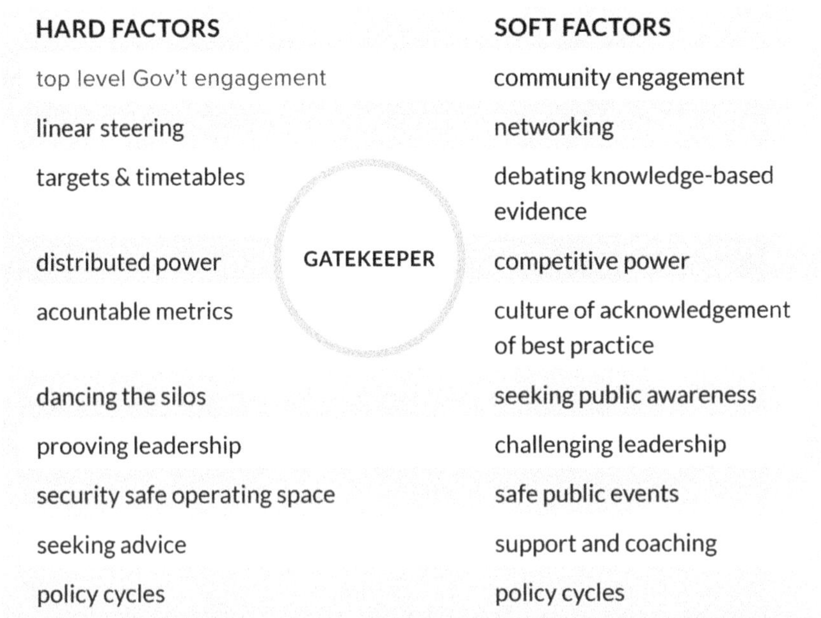

Figure 6.1 Governance context

Governance ideas can be great if they clear away the concrete obstacles and open up new opportunities. But they can also be grandiose and overblown, for example, when complex and wordy mechanisms are developed to deal with fairly straightforward problems. Sometimes governance issues are ultimately just an attribute of marketing and a courting of attention. Then they bluff eloquently and blusteringly in order to ultimately push through profane construction projects, products or lobby interests and make opposition appear illegitimate.

Governance in the sustainability context mixes hard and soft factors, where the hard factors concern administrative action, while the soft factors capture the social and cultural context. For beginners in this field, it is important to know that no strategy and not even the mature strategies contain all elements. There is no point in striving for perfection.

National and regional conditions must be taken into account, of course. Germany, for example, has a federal structure. This means that power is exercised both centrally and at the regional (Länder) and local community levels. Reaching multi-voiced decisions is sometimes costly and may take a long time. On the positive side, decentralised power systems are more likely to encourage more people to participate actively. New procedures are tried out and gradually enter the political culture. Only recently, good experiences have been made with some sort of citizens' councils made up of people selected on a representative basis (not according to prior knowledge or expertise) to deal with issues such as democracy, climate protection, etc.

Obviously, any governance scheme has to deal with the problematic of possible or practical greenwashing. No one can rule out the fact that nowadays even those, among countries and business leaders, who have not earned it honestly sometimes talk about their sustainability: The German stock exchange listed company Wirecard is a recent example. This company professed sustainability and publicly exhibited its corporate responsibility while manipulating its balance sheet on a grand scale. The exponents of the German meat industry are present on the internet with professional sustainability reports that stood in stark contrast to the poor social dimension of sustainability leading to large Covid-19 outbreaks among workers. The German automotive industry is now clearly focusing on sustainability solutions after decades of dodging the issue. But unlike all the "dernier cri" e-mobility fleets, sports utility vehicles with an unladen weight of 2.5 tons and 600 horsepower are still city tanks today, driving past the future of urban mobility.

The food industry is another example. The 2020/2021 lockdown increased retail food sales in Germany. This is not surprising due to the closure of restaurants and canteens. However, it is significant that organic food is purchased disproportionately more than before. The total value of the organic market in 2020 was around 15 billion euros and increased by 22 percent in 2020. The industry took that as business-as-usual. They waited in vain for a general offensive with regard to sustainable consumption, which might well have been expected in view of this growth. It would also have been more than necessary in terms of public health and individual anti-Covid-19 resilience.

Also, less serious cases sometimes declare something as sustainable too superficially, which does not stand up to closer consideration. It is too easy to come up with formulas according to which one only has to "appear sustainable" in order to be creative. Those who act in this way only conceal a deep creative coma.

The more sustainability becomes a project for everyone, the greater the demands and expectations on leadership, responsibility and (sustainability) strategic performance on the part of governments, businesses and the leaders of civil society organisations. The greater are also the demands for coaching and creative choice editing. Transformation does not exist as a one-off and in one size fits all (see Figure 6.2). No grand narrative is possible without differentiation and without out-of-the-box margins, without the diversity of actors, without different speeds and without centre stages and sideshows. No one likes to subsume himself/herself under planetary boundaries, world conscience, or a singular great transformation for example, if he/she thereby dwarfs himself/herself and makes him/her an object instead of being able to act as a subject. This is true for individuals as well as for groups and institutions. For them, these rough scaling often mean an underachievement. What is perceived with them does not exhaust what is perceivable, with all its unknowns and all its cultural features.

The transformative practise is a surprising and adventurous feedback loop. When it is not surprising and running in a loop, it is not transformative.

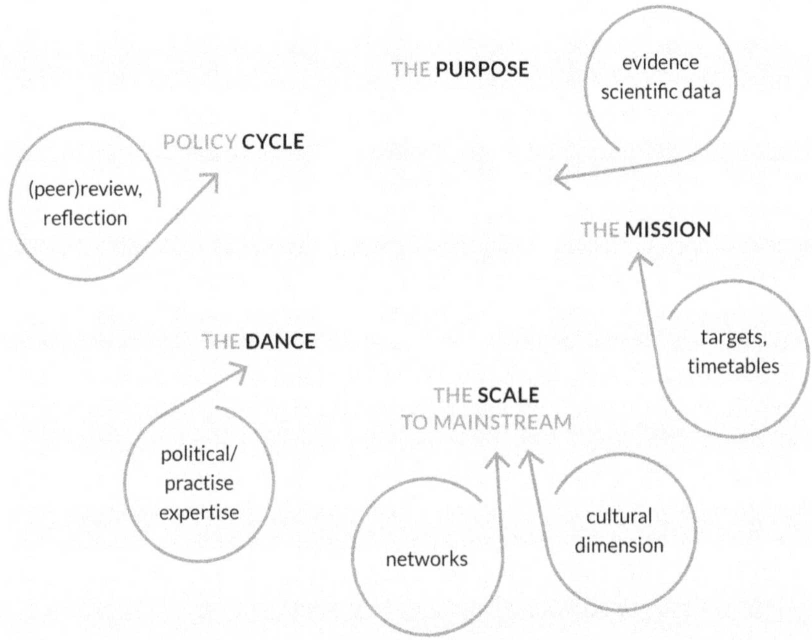

Figure 6.2 Transformative practise

For beginners and fast followers, it may be important to know that you can enter the loop at any of the five stations. The most common, of course, is to enter via the "Purpose" station. Driven by scientific evidence and triggered by concern about the state of things or the future prospects of unsustainable trends, people find this purpose. A purpose is usually something that can be retrieved from normal life, meaning it is already there and does not need to be invented or created. From there, you can start thinking about goals and time lines to develop a mission, including assigning the gatekeeper role. After including aspects related to the cultural dimension, expanding and reaching the mainstream of society and business would be an option. However, it is not the case that a strategy simply lines up all its active players in a linear fashion. The practise of alignment and integration might work for administrations, if anything, but definitely not in the social context. The station that is to come now is therefore called The Dance. The Dance is a good symbol for the challenge of overcoming the paradox of subtraction and addition. It is a phase of stopping and going, of giving and taking, if impulses and dialogue, working the silos and networking, without ever losing the big picture. The fifth station, obviously, deals with the construction of the policy cycle. Such a cycle is never a given, as has to be rebuilt again and again to keep the issue of sustainability from fading away. Organisational learning, and peer reviewing practices in particular, help.

6.4 The culture of multistakeholder engagement

At all of the previously mentioned stations, the involvement of multistakeholder initiatives (MSIs) promises to increase effectiveness. In the German context, the Sustainability Code (DNK), the Regional Network Hubs for Sustainability Strategies and the international exchange of experience, RENN, and the "The German Sustainability Award" (GSA) are particularly noteworthy here.

In MSI, actors from different backgrounds come together voluntarily and intentionally to solve controversial problems. MSIs are therefore defined by a commonly agreed process, rather than a prescriptive position on solutions. This distinguishes them from alliances and campaigns as they are common in civil society and association politics and are sometimes also initiated by states.

Globally, the sustainability agenda relies on multistakeholder partnerships. MSIs have so far played a much stronger role in global development policy and in Anglo-Saxon discussions on social responsibility (Hemmati 2020; Dodds 2019; Strandenaes 2018) than for the social market economy in the "association state" of Germany. The German government, namely the ministry for development cooperation, also refers to the various forums, dialogues or due diligence initiatives. They are about certifiable social or ecological standards for products such as textiles, palm oil, cocoa, diamonds. Cornerstones for MSIs, according to my practitioners' experience from Germany, are as follows:

- Legitimation:

  MSIs always need renewed legitimation. There are different ways to do this. (a) They can receive legitimacy by commission or legal mandate. (b) Legitimacy can be attributed to an MSI by third parties. (c) Legitimacy can be acquired in the course of the work and through a good design of an MSI, such as through results or/and through the process of the work.

  Mixed forms are also possible. The Sustainability Award falls into the third category, most MSIs that agree on standard for products or production methods fall into the first category. In any case, this legitimacy must be expressed and communicated in a comprehensible way.

- Discourse mode:

  MSIs are usually characterised by an explicit functional mode with which they deliberate their issues. (a) Their discourse is well-prepared and fact-based, which is imperative when the dialogue is triggered by different starting positions and hypotheses. (b) They acknowledge the respective practical as well as normative frame of reference of their positions and bindingly agree on the forms in which they want to clarify the open issues. (c) MSIs define whether and how they will keep their deliberations and work open or closed to the wider public.

  Of course, it is best if the mode of discourse allows the opposing participants to present and explain the respective counter-position in such a way that the respective cited side can agree. What would even top this highlight is when both sides would then check what they mutually might have failed to consider.

- Compliance:

Ideally, the initiators and participants take over the financing of their own initiative and create a legal entity with operational capacities. They also separate responsibility for strategic management from operational management of the available financial resources through contractually bound delegation or "Chinese walls" arrangements.

In practice, however, this is rarely the case. Often, MSI funding is primarily provided by the government. Therefore, the participation of civil society organisations and representatives from the private sector is undeniably a governance issue.

- Self-limitation and self-efficacy:

MSIs are not entered into for perpetuity, but for a specific time or occasion. The parties involved should always remember this purpose and the planned end of the cooperation. At best, they are time-sensitive and know exactly what policy cycle is appropriate for political/economic/societal decisions they want to focus on and influence.

- Context and impact:

MSIs turn complexity into a strength and a means of dealing with wicked problems that no single stakeholder group or no single line administration can solve alone. MSIs increase diverse expertise, and participants share competencies that are collectively greater than those that any one individual possesses. This is typically needed for successfully dealing with some environmental issues and global good policy issues, such as the production of palm oil and other commodities, ocean policies, or inclusive green economies.

MSIs can increase the quality and implementability of decisions and increase the chances of their impact, but this is not a given. There are also risks or possible shortcomings. Concluding from practical cases the renown MSI organiser Minu Hemmati (2020) lists e.g. unduly increased influence of special interests, influence of biased interests, lack of transparency, cost risks. Critical of globalisation, MSI Integrity (2020) claims that all (!) stakeholder initiatives and platforms miss their targets. According to this source, the 40 or so most significant initiatives did not meet their social and environmental goals and stood in the way of setting regulatory sanctioned obligations.

German environmental and social associations also tend to view MSIs sceptically but more constructively when they look at the power mechanism of MSIs. They basically argue that MSIs merely circumvent legal regulation and thus undermine the binding force that is actually needed. Also, the investment of time and resources seems disproportionate to the effects (Venro et al 2020). The core issue here is also how voluntary participation is, respectively whether and how its refusal and misconduct are sanctioned.

However, regulatory measures on the wicked problems. tipping points, and runaway problems require that the stakeholders concerned bundle their approaches, enter into cooperations and have an impact beyond the market influence of a respective individual company into the industry (competitors) and into the supply chain (upstream and downstream). However, the opportunities for cooperation in the area of sustainability also encounter barriers, including those arising from antitrust law, and bottlenecks arising from attitudes, habits, and positional politics.

It is questionable whether MSIs will become more relevant and effective as long as their legitimacy is seen merely as a kind of substitute for state regulation. In the future, the legitimacy of an MSI is likely to be measured primarily by how well and bindingly the MSI works as an interface platform, and how well this specific governance is anchored. Properly applied and developed, they would contribute to a new and powerful policy paradigm for sustainability:

The Dutch Sustainable Trade Initiative offers an interesting example of better, because cross-cutting, governance. As a cross-sector and cross-product initiative, it brings together manufacturing companies, governments and the financial industry (see IDH (n.d.)).

The founder and executive chairman of the World Economic Forum Klaus Schwab (2020) prominently affirms that neoliberal ideology is wrong and that what is needed now is a paradigmatic shift to what Schwab calls stakeholder capitalism, which requires open data approaches and, in the long run, the creation of a sustainability cloud.

High-performing companies are already supplementing corporate management with consideration of benefits and losses from ecology and human rights along the supply chain which, as non-financial or still externalised aspects, do not appear on the balance sheet. In the next few years, this change would begin on a voluntary basis, but will soon have to merge into a corporate – public system of standards and regulations, or will fail altogether. The technical progress around digitality and data ecosystems is already there. The data economy makes it possible. Carbon metrics, and in other cases: metrics covering the nutrition score (sugar), full loop recycling of materials, or biodiversity impacts, are requiring up-to-date data infrastructures, including the visualisation of big data. Corporations, for practical reasons, are required to build and service their own sustainability clouds.

The new data ecosystems consist of oceans of granular individual data that are intrinsically worthless. They obtain their value through the assignment of attributes and algorithmic processing. Attributes characterise evaluative variables, benchmarks, and standards; however, these numbers are more than just counting variables because they must express ethical values and cultural practices in order to be socially recognised. In other words, they require dialogue. It is a huge task. It can be pursued in parallel by each government institution on its own and by each company to the exclusion of its competitors. That would be the usual, but not appropriate, way. It would run into limitations, and would threaten to lead into data bureaucracy and double counting, merely because supply chains are interwoven and complex. The alternative at hand is the development of a common platform solutions, in the sense of a, say, European Public Data Space on Solutions for Sustainability. There is already a practical reference in Germany, namely the MSI Industry 4.0, which is driving forward the digitalisation of robotics. Clearly, MSIs should play a role in this context that is hardly comparable to today's.

6.5 The benefit of a multistakeholder advisory body

When a government (or other legal entity) decides to establish a multistakeholder body, it sends an important signal to the public. It is a sign that sustainability is taken seriously. The political force of this signal will be stronger the sooner the following four technical criteria, which represent a trusting political culture, are realised or come close to them. The (a) endowment of the body with meaningful resources is the first critical factor, followed by (b) a reliable and confidential channel of communication between the body and top government authorities, (c) the right of the body to pursue an independent agenda, (d) the ability to implement projects that reach society at large.

Although it might stand for itself but it needs mentioning: the mandate of this kind of advisory body must be tied to the national sustainable development policies. This tie legitimates the body more than the original appointment does, because this can be merge into a living tie that renews, strengthens or reinforces the good practise of producing advise. Another underlying feature that nonetheless needs to be highlighted is the positioning of the body in the centre of the government. For the central positioning there are several models. Finland decided that its multistakeholder advisory body is being chaired by the Prime Minister himself, and a number of ministers are regular members of the board, next to eminent societal leaders. This construction has a number of benefits, such as the direct line of feedback and access to implementation, or the availability of more time for internal pro and contra debates. In the German case a chair runs the body that its members are electing from among the membership in secret voting. Meetings with top political leaders of government are by invitation only and must be mutually agreed upon, which means they are never routine and issues seldom just fade away. Attendance at meetings of the high-level decision-making body of Secretaries of State, on the other hand, is a standard format.

The Finnish and German options are representative of a wide range of governance options implemented in other countries, i.e. when the country's government feels encouraged by the political culture to establish an advisory function in the first place. Both options have their pros and cons in terms of transparency, agenda and impact, debate and prior-consensus conflict management. Both concepts, of course, enable multistakeholder bodies to develop networks and engage in dialogue with other stakeholders (see Figure 6.3).

Already the size of the board signals a statement. Basically, one can differ two types of multistakeholder bodies. A smaller number of members allows appointments to focus on the personal skills and reputation of the members in an ad

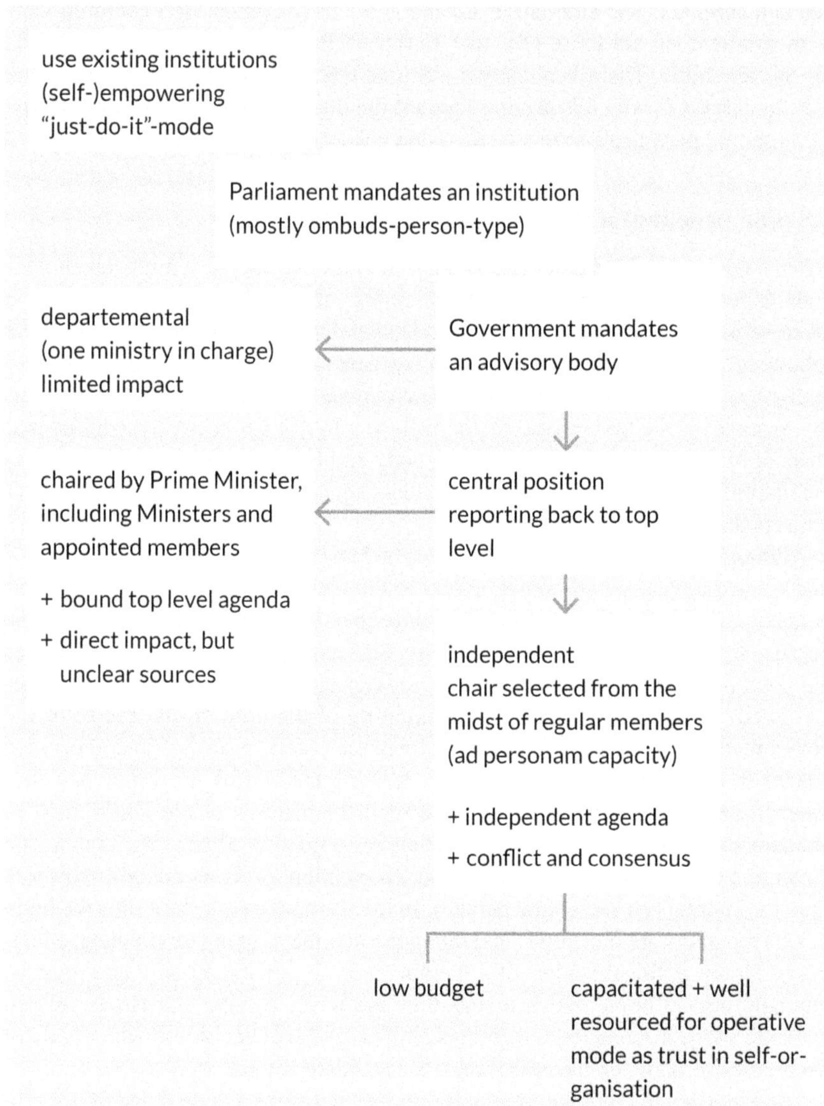

Figure 6.3 Alternative options for establishing gatekeepers

personam style. The number of members may not exceed 20 if the body is expected to conduct ad hoc deliberations and to meet political deadlines. The number of members can be much higher if the body is to formally represent all relevant sociotopes of a given society along almost parliamentary lines. The government could ask organisations to send members to the body. Both types of bodies can be of value, the first as an advisor and facilitator, the second mainly as a forum for discussion and exchange of opinions. However, a mixture of these types is unsuitable for both purposes.

Another option is to build an advisory upon pre-existing institutions. This means re-organising responsibilities and may contradict with the self-image of given institutions. A further option is the strengthening of subnational networks and of local engagement of citizen groups, the importance of which must never be neglected but, again, will not come close to the benefits of a regular multistakeholder bodies with privileged access to the government.

Those benefits are manifold. A sustainability advisory body (council, commission, board) might be able to reach a consensus about a topic that is highly controversial among the political partners forming a government. So, consensus building would be the impact of the body. Similarly, projects could open doors in society and can build trust among parts of society that so far have stayed away from any engagement in sustainability efforts. Convening debates and facilitating the review work of international peers might be another facet of what a multistakeholder body can do. With facilitating feedback mechanisms it would add value to national politics that otherwise cannot be achieved. Last but not least the body – through its appointment, central positioning, and privileged internal communication to the top governments levels – will create space for action, e.g. by providing a well seen stage for public debate, by encouraging the creating of funds and programmes and by informally encouraging leadership across societal sectors and sociotopes that otherwise seldom come together.

On this basis, some general conclusions can be drawn for those who consider establishing a multistakeholder SDG advisory body. Those conclusions may also provide some answers to often-asked questions for existing multistakeholder groups wanting to play a more significant role.

First, all this is about political culture and must clearly be beyond wordy rhetoric. Conversation without depth and self-reflection is ultimately just boastful communication and driven by ideology. The usual political rhetoric can rarely be relied upon. What is needed is a political culture beyond ideology: listening attentively, exchanging ideas without prior bias, learning from other views. Differences should not be blurred in the process; rather, they can be introduced in a principled manner and on the basis of ethical values. Patience is a resource, as is a certain humility toward the Anthropocenic task.

Second, comprehensive leadership counts. For a multistakeholder group to be effective and successful, leadership of the group is crucial. Characteristically, the groups leadership must be empathetic, but at times must also act with authority. It works best in close conjunction with the professional staffer of the group, who in turn maintains trustful links with policy and other addressees, and who holds a political mandate to represent the panel. His most appropriate working modality is that of an agent and innovator, in addition to that of an executive servant. Internally, the leader gives members a sense of a common search for new approaches, and externally, that the cycles of action under which policy operates are understood to the extent that meaningful recommendations can be expected.

Government appoints multistakeholder

→ 15 members + 1 Secretary General
→ 3 years terms
→ independent agenda
→ budgeted + staffed

Involving Government by giving advise, providing a safe space for public debate, initiating societal action, addressing the private sector

Figure 6.4 Modality of RNE

Such panels discuss the pros and cons of inaction and action and look at the whole panorama of prosperity, future, hope, financial and environmental responsibility, economic interests, and engineering competence.

Third, numbers matter. With the total number of appointed members, the government (or other appointing body) sends a strong message. When members are appointed on an "ad personam" model, i.e. based on their personal expertise and particular bargaining power, a smaller number of members signals high expectations for the commission's outcome. It promises tangible results. With a limited number, the government demonstrates organised and thoughtful expectation management. The same is true for staffing and budgeting the commission. There is also a rationale for larger commissions. Given the specific political culture of a particular country, it may make sense to appoint members according to rules of formal representation. Coverage of regional constituencies and/or indigenous peoples may necessitate a larger number of members, leading to a more parliamentary working model (see Figure 6.4).

Fourth, the virtue of clarifying conflicts. How members of such groups handle their controversial positions and interests is of paramount importance. It is not advisable to seek only the lowest common denominator. Simply announcing that you are looking for the lowest common denominator leads to it becoming smaller and smaller. On the contrary, opposing viewpoints need to be made clear. The repetition of these controversies can be useful for members to feel accepted and to distinguish themselves. In the best case, conflicting parties can restate the points of the respective opposing party in such a way that the latter feels understood and accepts the presentation. However, this alone does not solve problems. The pure confrontation of A versus B may overlook the possibility of a C or a D. Searching for the overlooked, the new, and the surprising might be part of a solution. A good consensus wins through well managed conflicts.

Fifth, soft rules create trust. Self-set soft rules may structure dialogue by manifesting the respect and dignity of the people involved. Of course, every discourse needs formal rules. The general one is that nothing is decided until everything is decided. This prevents the premature driving of stakes on individual issues that may prejudice or devalue further negotiations. A suitable working model is the so-called living document. One person is in the position of the chief editor of the document, while all participants can contribute texts and comments at any time, which the editor consolidates into a living document after consultation with the committee chair. This procedure begins with the first scoping and ends with the completion of the recommendation. The benefit is that all step-by-step versions of the document are always discussed in its overall context. This procedure serves to build trust and is particularly advantageous under time pressure. The interested public should be informed appropriately.

6.6 Change of culture

Never before are more people in Germany committed to not throwing away food, recycling waste, taking to the streets for healthy food or for effective climate protection and an intact environment, to fair trade and respectful interaction. Never before have such large segments of the business community embraced sustainability as their mission or part of their purpose. Never before have politicians set more ambitious goals for sustainability. What's more, humanity's intellectual and material resources and opportunities are more extensive and diverse than ever before. Victory over world hunger would be possible, a stop of global warming would be possible. It would be possible to stem the tide of waste. It would be possible to make access to education fair. It would even be conceivable to make the number one and number two pollutants, carbon dioxide and methane, a valuable resource – instead of dumping them in the atmosphere and the oceans. Even a more accurate sense of responsibility about what constitutes economics and what nature means for economic activity would be possible.

The change in cultural awareness speaks to this. Consumers are concerned with the issue of sustainability, which can also be observed in the fact that sustainability plays a major role in commercial advertising today in the food and consumer goods industry, the capital market and the fashion industry, and finally the automotive industry. Just a few years ago, the mere word sustainability was considered an advertising killer, a no-go for marketing, and major German print media used it only in quotation marks. Today, the market power of sustainably operating companies is still growing. They learned that a focus on sustainability can be part of the "licence to operate" which is non-regulatory but not less impactful; and it can bring advantages, such as greater efficiency, the launch of reshaped or new products, a better reputation, satisfied employees, hiring power, higher qualification of young employees, meeting demands of investors and business-to-business partners. There are also advantages to be gained by recognising early on where the next innovations or risks lie.

Thus, the times are over when sustainability could be spitefully ignored or benignly postponed until tomorrow. On the other hand, in the German spectrum of opinion, there is a minority that rejects sustainability as a mere fashion trend and as hysteria. They hope it will soon disappear. One also often hears the accusation that politicians are only simulating sustainability and that nothing is actually changing. Voices deny the man-made contribution to climate change and any other scientific evidence, and they consider the phasing out of nuclear energy to be the devil's work. For them, sustainability is an aberration. These voices are mainly confined to the special interest groups in the echo chambers of the internet, and since they are not heard by the general public, they seem like the sleeper department of non-sustainability. That doesn't make them any less dangerous. What they actually express is a current form of German Angst. The critics look with horror at the transformations already underway and fear that they will no longer be able to keep up. It is a fear of losing control. People always fight against these feelings in a fear-fixated and particularly dogged manner.

The common sustainability discourse may also play and cause some of this vehement rejection. It is often too fixated on the nomenclature of associations and actors, on the state, the law and also on the next programme of state funding. It either promotes a first-come, first-served approach to innovation and competitiveness, or it promotes a nanny approach to the role of the state and its administrations. It also flirts too much with the idea of command centre on moral high ground that would force change from the top down. Too often, the discourse seeks refuge in apocalyptic mirages and loads (even correct) accusations of guilt all too heavily with the moral index finger. Too seldom are conflicting goals and contradictions openly addressed. Far too rarely is the cultural dimension actually fleshed out beyond wordy claims.

In 2020/2021, Covid-19 teaches us the hard way that precaution is essential for survival, and that – a cultural principle – our behavior as individuals not only determines our own health, but also the lives and health of others. The lesson learned from the pandemic is a prime example of what is meant by the cultural dimension: (a) The scientific-industrial breakthrough for the fast production of a vaccine is only conceivable at all as the result of a cultural environment that promotes basic research, tremendous scientific excellence and socially responsible entrepreneurial initiative. (b) Technical-medical assistance and infrastructure must go hand in hand with the behavioural culture of individuals to effectively combat a pandemic. (c) Political decision-making in Germany was evidence based and has long openly addressed its knowledge base, including uncertainties, with full pressure to act, thus gaining credibility. (d) Averting danger and saving lives had to be coupled with precaution and prevention. (e) The precautionary protection of those at risk was carefully weighed against restrictions on civil rights.

Precaution, both environmental and public health, carries a paradox. If precaution works well people tend to forget about it and politicians think that precaution is an exaggeration and a pretext for non-action and an excuse to prevent progress from happening. This, however, is not the case. As a matter of fact, any anticipatory responsibility and precaution are difficult to calculate, and disputes must arise about whether they are necessary, what they may cost, who should be involved, what probative value data have. These arguments must be seriously addressed. Prevention is the economic principle of doing the right thing at the right time. This is the deep truth of climate change as well of a pandemic caused by zoonotic disease overshoot. Both cases demonstrate that delaying precautionary action makes the solution more expensive. Delay is aggravation.

Thinking ahead helps, as does reassurance about what has been achieved, about successes and mistakes, and the cultural patterns that are interwoven with them. This is often the source of creativity.

References and recommended reading

Brouwer, H., J. Woodhill, M. Hemmati, K. Verhoosel and S. van Vugt (2015) *The MSP Guide – How to Design and Facilitate Multi-Stakeholder Partnerships*, www.mspguide.org/msp-guide

Cordonier Segger, M., M. Szabó and A. Harrington (eds.) (2021) *Intergenerational Justice in Sustainable Development Treaty Implementation: Advancing Future Generations Rights Through National Institutions (Treaty Implementation for Sustainable Development)*. Cambridge: Cambridge University Press.

Dodds, Felix (2019) *Stakeholder Democracy: Represented Democracy in a Time of Fear*. London: Routledge.

Gebhardt, Beate (2016) *Beschreibung von 24 Nachhaltigkeitspreisen in Deutschland mit Relevanz für Unternehmen der Ernährungsbranche*. Institut für Agrarpolitik und Landwirtschaftliche Marktlehre der Uni Hohenheim; Arbeitsbericht No. 26, https://marktlehre.uni-hohenheim.de/fileadmin/einrichtungen/marktlehre/Arbeitsberichte/haa26_net.pdf

Gebhardt, Beate (2017) *Bedeutung von Nachhaltigkeitswettbewerben für Unternehmen in Deutschland. Studienergebnisse 2017 [Relevance of sustainability competitions for companies in Germany. Study 2017]*, https://doi.org/10.13140/RG.2.2.31904.84485; www.researchgate.net/publication/332222351_Bedeutung_von_Nachhaltigkeitswettbewerben_fur_Unternehmen_in_Deutschland_Studienergebnisse_2017_Relevance_of_sustainability_competitions_for_companies_in_Germany_Study_2017

Gebhardt, Beate (2020) Quo vadis Nachhaltigkeitswettbewerb – Bedeutungsverlust als existenzielle Herausforderung? *Konferenz: Möglichkeiten und Ansätze zur Qualitätssicherung von Nachhaltigkeitsawards für Unternehmen*, https://doi.org/10.13140/RG.2.2.32075.21286; www.researchgate.net/publication/345601752_Quo_vadis_Nachhaltigkeitswettbewerb_-_Bedeutungsverlust_als_existenzielle_Herausforderung

Hemmati, Minu (2020) Stakeholder Engagement: With contribution of B. Maier. In: M. Paschke and M. Dahinden (eds.), *Engaging in the Science-Policy Dialogue*, Workbook 2. Zurich: Zurich-Basel Plant Science Center, https://doi.org/10.3929/ethz-b-000308464 (open access)

IDH (n.d.) The Sustainable Trade Initiative, the Netherlands, www.idhsustainabletrade.com/about-idh2/

Kannegießer, Matthias (2017) *Zehn Jahre deutscher Nachhaltigkeitspreis. Wirkungen und Perspektiven für Nachhaltigkeit in Unternehmen*. Berlin: Studie im Auftrag des Rates für Nachhaltige Entwicklung, www.nachhaltigkeitspreis.de/media/7-Presse/Downloads-Allgemein/studie-10-jahre-deutscher-nachhaltigkeitspreis.pdf

MSI, Institute for Multi-Stakeholder Initiative Integrity (2020) *Not Fit-for-Purpose: The Grand Experiment of Multi-Stakeholder Initiatives in Corporate Accountability, Human Rights and Global Governance (Summary Report)*, July, www.msi-integrity.org/wp-content/uploads/2020/07/MSI_SUMMARY_REPORT.FORWEBSITE.FINAL_.pdf www.msi-integrity.org/not-fit-for-purpose/

Schwab, Klaus (2020) A Better Economy Is Possible: But We Need to Reimagine Capitalism to Do It. *Essay in der Ausgabe the Great Reset des Time Magazine*, October, https://time.com/collection/great-reset/

Strandenaes, Jan-Gustav (2018) *Towards Strong and Active Stakeholder Engagement at the Global Level.* Presentation at the 2018 Sustainable Development Transition Forum Accelerating Progress Towards the SDGs: Enhancing the Role of the High-Level Political Forum (HLPF), UN OSD, Incheon Songdo, Incheon, Republic of Korea, https://sustainabledevelopment.un.org/content/unosd/documents/4884Session%205-1_Jan-Gustav%20Strandenaes.pdf

VENRO, Cora, Forum Menschenrechte (2020) *Anforderungen an wirkungsvolle Multi-Stakeholder-Initiativen zur Stärkung unternehmerischer Sorgfaltspflichten.* Empfehlungen aus Sicht der Zivilgesellschaft, https://venro.org/publikationen/detail/anforderungen-an-wirkungsvolle-multi-stakeholder-initiativen-zur-staerkung-unternehmerischer-sorgfaltspflichten

7 Failures

An indispensable resource

7.1 Failure culture

Failure is in the nature of doing things, in dealing with injustice and the forces of nature. The question is not whether one makes a mistake or avoids it. The question is what kind of mistake it is and whether you understand its lessons. If failure is the most valuable resource for correcting oneself and setting out on the path to success, then, why are failures the facts least openly discussed and publicised by policymakers, as well as by stakeholders around the topic of sustainable development?

Failures happen a lot in the world of sustainability. One could even say that the whole theory of sustainable development is built on the missteps that real development has taken so far. But that would only be rhetoric. More important are the practical mistakes that occur in the context of positively intended, political responses to problems. Examples are manifold: politics underachieve binding goals and miss out on opportunities; an important technical solution, though on the table, is not addressed until decades too late; stakeholders fail to reach a compromise for a common demand; administrative governance deepens gridlock instead of cleaning it up; climate goals are not met, although public support could hardly be greater; the circular economy is not moving forward, despite the trillions that the consulting firms never tire of saying are on the table; and much more.

For example, in the case of combatting global warming the most serious mistakes occur not because of what we do, but because of what we do not dare to do or of what do not even think about, at the right time. Not in all cases, obviously, but mostly errors of omission are involved, and thus some errors are hidden and less visible than others. In operations that follow predictable rules failures may be prevented; in complex systems they are unavoidable and may have been years in the making. In the Anthropocene the overall substance of failures is time. Time is a mediator of justice, as already expressed by the definition of sustainability given by the Brundtland Commission. Positively formulated, therefore, a vision with execution is one of the best drivers of success, and, on the contrary, the absence of vision and execution is an equation for failure.

Dumb failures are dull failures. Smart failures are a value proposition. They provide insights that can lead to breakthrough benefits. Similar to the usual return

DOI: 10.4324/9781003242024-7

on investment, there is a return on failures, but this one is extraordinary, and not a mere function of money. Smart failures may arise from deviating from the usually assumed course of action, breaking new ground, or checking out alternatives involves all kinds of possibilities to run into failures. In the private sector, and especially in the startup community, the proverbial culture of failure is definitely considered sexy. Fail fast – fail often – fail forward is a well-known mantra in this community. Business students learn that purpose, trust/confidence, and connection are the key principles of emotional intelligence, which enables systematic awareness of risks and opportunities.

You won't hear any of that in public administrations. There is zero explicit failure culture in public administration. For ministers and administrative leaders, failure is a hard word to say. In public politics in Germany, admitting errors comes next to resigning from office.

Generally, while demanding a failure culture is an often-heard mantra, the practise of extracting lessons from failures can still be a walk on the highline and may be emotionally exhausting. Theorising failure is not what is needed. Applying concrete trainings and a coaching offer would be fair next step.

The German sustainable development strategy has clarity as to whether or not their goals have or will been successful. Statistics and indicator speak a clear language about targets that are on-track and will probably been met, and off-track targets where the real development is dodging the binding target. However, even solid data do rarely provide the framing for a deep understanding about the nature of failure and its conditionalities. Paradoxically, political assessment of off-track situations often reinforces well-known charges of the usual suspects while failing to examine the root causes of persistent failures, or, even worse, the next action is replaced with theorising on things that could possibly go wrong in due course.

7.2 The global machinery sputters

The HLPF, the United Nations High-Level Political Forum, is the highest universal diplomatic institution dealing with the state-of-the-art of sustainability. It oversees the process of implementing the Sustainable Development Goals. The today's online social media parlance compares the HLPF to an app that is used to communicate and that has a messenger service attached to it. Along with other networking sources, it can be used to figure out how where the world stands in delivering against the 2015 agreed goals. As an institution the HLPF corresponds with other processes that have different settings. To name a few, the conferences of the parties for climate change, or biodiversity, and the finance for development conference are the most important ones.

HLPF and the respective UN bodies, such as UNDESA, deliver thematic reviews and scientific reports. Informal explorative meetings and laboratories add value to the international conversation. Registering voluntary national reports (VNRs) on how states are adapting to the SDGs the HLPF delivers a horizon scan of nearly all UN member states. Sometimes those reports hide more than they describe, but there are also such reports that show a set of different perspectives, such as those of the state, interest groups, and the business community. Inviting stakeholder to have their say in itself gives a more comprehensive impression and promotes self-reflection.

As for the European context the European Sustainable Development Network (ESDN) gathers representatives of the public administrations of European countries responsible for sustainability and other experts. ESDN summarises the state-of-the-art of sustainability strategies in many reports and expert events and organises the European Sustainability Week. The network of European national advisory bodies on the environment and on sustainability (EEAC) is linking national independent expertise from those countries where such bodies exist (Niestroy 2005; de Vries 2019), which as of now it the case in Belgium, France, Germany, Hungary, Ireland, Luxembourg, Montenegro, The Netherlands, and Portugal, with subnational Councils from Catalonia and Flanders. At the global level, the UN Office for Sustainable Development, UNOSD, in Incheon, Republic of Korea, has a good overview of national implementation of the SDGs. UNOSD is an operational institution that facilitates the sharing of national experiences. Also on offer are training and coaching, as well as expert advice to help build technical and human capacity, and, thus, fills the gap in knowledge management on the part of administrators and experts in developing countries.

Neither institution is known for systematically criticising its own failures to implement sustainability or the failures of national governments. Failure culture is not a quality at home in the multilateral system which is quite understandable when it is not even present at the member state level. In an ideal world one might assume that multilateral institutions should lead the way and set an example here, as they are far removed from the most immediate demands for "street cred" expressed in polls, votes, and elections. This is, however, not the case. On the contrary, it is rather blocked by budgetary turf battles and positional claims that make any criticism appear simply hypocrisy or foul play or something completely unpolite and basically illegitimate. The only voice of criticism is that of the UN Secretary General Antonio Guterres. In September 2021, he used a clear language when he spoke about systematic and manifest failures. Introducing his report to the General Assembly (Guterres 2021) he compared Covid-19 as a wake-up call for global responsibility, true also in the fight against global warming, and he observed, he said, that everyone is oversleeping in some type of paralysis: "From the climate crisis to our suicidal war on nature and the collapse of biodiversity, our global response has been too little, too late . . . Multilateralism is failing its most basic test".

The multilateral decision-making system may not be overly modern or effective, but it is there, at least in its rudimentary form, and it should be held accountable. That should not be forgotten, and fundamental organisational reforms have long been needed. Guterres' harsh criticism is correct, yet it is just another symbol of the complacency of world leaders, who frequently lament urgent goals and implementation gaps without consequences that even halfway match their dystopian language. A criticism from the very top (Guterres), as correct and personally credible as it is, is part of the problem if it only triggers a collective chorus of larmoyant agreement, and a sum of individual actions that run counter to the chorus.

So, quite a few political leaders see for their countries more immediate needs than sustainability and global goods. That equals the action of business leaders who care about growth and ebit more than about anything else. Voices from the far right call the 2030 Agenda a remnant of an army of poor leftist agents. As absurd as that sounds, it is part of the real world, and we have to deal with this.

Fortunately, many countries report a high importance they give to the 2030 Agenda. A functioning public sector and a responsive society may find approaches that differ from the usual routines. Governments may choose from direct action, parliamentary resolutions and the sanctioning mechanisms of auditing. (The International Organization of Supreme Audit Institutions, INTOSAI, could even expand its role in reviewing compliance and progress made.) They may also put more weight on the community level, or may prefer passing pieces of legislation that establish a bill for the rights of future generations like in Wales. Establishing ombudspersons for future generations and other binding constitutional arrangements give an idea of complimentary legal options (Szabó 2021). In countries with engaged civil society and enterprises those are running pioneering initiatives. The Italian Alliance for Sustainable Development (ASviS), for instance, gathers organisations in the economic and social field. With a large-scale festival these alliances raises awareness and fosters cultural-political reflections on the SDGs. Herewith, the alliance sets points of reference for all activists, including the Government. In Nigeria, the Jesuit priest John Patrick Ngoyi, has built up the Justice, Development and Peace Commission into the centre of a myriad of advocacy initiatives of an active citizen sector that is promoting the national adaptation of the SDGs, an activity that the Nigeran voluntary report is not even officially mentioning or highlighting. It brings hope to the people that their lives, the economy and the human rights landscape will improve. The deceased King Bhumibol Adulyadej of Thailand applied sufficiency management practices across agriculture, business, government and community organisations (Avery and Bergsteiner 2016). The Thai sufficiency project might not have been the holy grail of sustainability (as which it actually had been introduced) but certainly it addresses a topic that is otherwise too often just omitted. The project Partners for Review, funded by the German Official Development Aid programme, is a global network that helps states preparing VNRs. This project works in conjuncture with the official HLPF process. In 2013 the Bertelsmann Foundation (2013) conducted a field research that covered sustainability politics in Bhutan, Costa Rica, Finland, Ghana, Tasmania.

Anyone interested in an update on the status of national sustainability contributions being presented and discussed at the UN cannot avoid CEPEI's reports and assessments. CEPEI is a think tank from the Global South based in Colombia. Its analysis and advisory work focuses on how the Global South in particular is positioned on the sustainability agenda, what role Big Data and digitalisation play (a central one), and how the governance of the UN, and here in particular the performance of the regional UN authority, can be improved (in the frame of a general organisational reform of the UN). The five UN Economic Commissions (Africa, Latin America and the Caribbean, Arab Countries, Asia Pacific and Europe – including Canada, Israel, the Russian Federation, the USA, Eurasia and Middle Asian Countries) conduct each a regional forum preparing meetings at the level of the UN headquarter. CEPEI's Governance for Development department monitors results by a series of infographics and reports on its website (CEPEI 2021).

All those platforms and fora, and in particular the HLPF, do not meet the requirements that policy makers have for educating themselves about their own options. The identification of what works elsewhere, and why, and whether a policy maker should try and import or adapt such solutions is a rather difficult task. The structure of the HLPF VNR exercise does not deliver that and is not made for that. Additional action is clearly needed. It should be able to convene even painful discussions without the countries' experts or the UN officials seeing that as disrespectful. This is a challenge not yet met, and on this level not yet even tried.

The German sustainability politics ventured a first attempt when the German Chancellor Angela Merkel invited high profiled elder statespersons and activists from around the world to come to Germany and advise her on how to advance sustainability policies. The members of those peer reviews, held three times in 2009, 2013, and 2018, were guaranteed an open door policy with access to ministerial departments at the level of their choice (experts, heads of units or of administration, up to ministers), parliament and stakeholders. The German example built on prior experiences of peer projects conducted by The Netherlands and France. Procedures that look into successes and failure, eliminating the dumb one's and learning from the intelligent one's is what the global machinery needs at scale.

7.3 Why evergreen governance goodies fail to deliver

Since the first "Earth Summit" in 1992 the policy framework of how to combine "environment" and "development" is subject to national decision-making. As agreed on UN level the national states are in charge of implementing the universe sustainable development goals, however, with no sanctioning of misconduct or ignorance. Even the disclosure of data and actions is voluntary. Since then there is no shortage of theories and wordy documents, but an urgent deficit in delivery. Now, the run-away-challenge of overheating the atmosphere and of species-die-off, sharpened by geopolitical instability, is increasing the sense of urgency.

How to implement long-term goals and do it faster and more reliably than ever before is therefore being discussed all over the world. Year after year, heads of state, ministers and top UN officials and top representatives of civil society complain at their regular meeting at United Nations headquarters that, while there have been small successes here and there, they have not yet met their own aspirations for sustainable development. Civil society groups agree. They too complain about being off-target and call for more political determination to change this. How to actually do this seems simple at first glance: more political will, better regulation, policy coherence plus integration plus whole-of-government coordination, participation of stakeholder, public access to monitoring and evaluation, clear alignment, and informing those who are still hesitant or resisting. These are governance "goodies". They are being discussed not exclusively as regards the sustainability agenda, but rather for all topics and challenges of modern societies. Now, however, they are failing to deliver. How else to explain the fact that almost nothing happens for decades, while political demands become more and more urgent? On the climate issue, Germany's political class is surprisingly clear: Almost everyone wants climate neutrality and wants it even faster than already decided. However, this "green race to zero" all too often remains at the level of targets. When a target is missed, or runs into an off-target situation the political mainstream reacts with an even more ambitious target. The how-to remains in question, likewise the question of governance.

The term governance refers to how public administrations and actors from private companies and civil society organisations relate to each other in order to achieve certain goals. It applies generally to all functions and tasks of administration. Governance is of particular interest for sustainable development because sustainability is a new field and because time and required action are of increasingly urgency that basically does not allow old style wait and see politics. Niestroy, Hege, Dirth and Zondervan (2019) analyse the diversity of governance tools in Europe's member states and in the European institutions where the trend is for governments to include the SDGs in their development plans or national sustainability strategies. Strategies are being developed into operational tools. The OECD (2019) suggests so called key governance pillars that the OECD experts expect to help governments to prioritise and set national targets and objectives across the policy spectrum. The OECD supports a shift from what it calls traditional siloed policy making to more integrated approaches. Again, those pillars are all about the evergreen-goodies of governance such as coordination, coherence, alignment, and integration.

So, what is it about governance? And why on the wish list are the very same evergreen goodies, again and again? Why is this not making a difference in reality?

- Coordination

The requirement for coordination relates dominantly to internal government action and administrative processes. Coordination is what continuously happens in an everyday routine of inter-ministerial meetings and technical processes. The point is: Everyone wants to coordinate, no one wants to be coordinated. It can be observed that a successful coordination with the "wrong" goal works as a stumbling block for new ideas and deviating goals. Not surprisingly, conventional give-and-take procedures prevail. The opposite, however, would be an open and mission driven coordination for which new dialogue formats are ever more important, but hard to administer. The oft-repeated mistake is to call for a fully coordinated policy in a whole-of-government approach. Rather, it is advisable to take the piecemeal approach and create and test a partial scope for task-oriented coordination.

Governmental decision-makers are often in need for a well-coordinated stakeholders' group in order to get full resonance on governmental measures and also in order to overcome hinderances. In their view, a conversation with stakeholders would make sense that would take place at a very early stage and obviously prior-to-decision-making times. In practise, that is a challenge. In order to engage seriously, and before being targeted in this way, stakeholders would feel obliged to ask for some guarantees, and they would want to know what they were getting into.

- Coherence

The coherence of government action stands very high on the wish list of any better-regulation-agenda, and of stakeholders and politicians who expect this to lead to the realisation of their own goals. According to the concept, a state acts coherently when all its parts are directed toward the same goals and its individual actions do not contradict each other. Critics of states lagging behind sustainability goals consider coherence of policy action to be a key requirement. They call for integration and alignment. Coherence strives for the ideal of a harmoniously functioning administration. However, this demand is often exaggerated and the demand for alignment alone leads to resistance because there are always some ministerial sociotopes misunderstanding it as self-sacrifice.

The demand for coherence does not understand the practical functionality in administrations and ministries. Those quite deliberately rely on the opposite of coherence, either because this is the best way to protect outdated behavior or because new procedures and goals cannot be implemented otherwise. In German government practice, it is common for the middle and higher levels of staff to make agreements and alliances among their peers (horizontally, if you will) in order to help their policy initiatives gain acceptance. They also do this across other administrations. New "silos" are continuously formed and old ones are rebuilt. Conflicts of objectives are fought out at this level, and only if this does not succeed does the next higher level of management become involved. Incoherent action therefore does not happen out of bad will or because of a narrow-minded tunnel vision, but can also be a deliberate result of functioning administrations. In this respect, a certain amount of deliberate incoherence can also create the prerequisite for demonstrating leadership.

- Integration of silos

This is similar to other demands are often raised by the organised civil society. For example, the existence of silo structures is deplored, where silos are political, mental, and institutional/administrative structures that function with a core set of attitudes and processes and defend them against influence from outside. There are strong calls for breaking down those silos that withstand, e.g. climate protection necessities. The alternative to breaking down fantasies is a better understanding of the internal dynamics of silos (they are continually reassembling) and of using a good silo for one's own purposes. Niestroy and Meuleman (2016) even suggest teaching silos to dance. Integrating administrative action, both vertical and horizontal, is also often called for in a very theoretical way. Ideal integration could only work in a singular, hierarchical structure of action, but in modern, differentiated societies this kind of hierarchy seldom works any more. Managing sustainable development means above all dealing with conflicting goals, achieving a fair exchange of burdens and benefits, establishing a credible and reliable flow of information, and changing mindsets and preconceived notions. The concept of integration is rarely helpful in this process; on the contrary, it tends to arouse fears that exacerbate the challenge of transition.

- Cooperation

While different streams of coherence compete with each other, cooperation is a serious practice. Cooperation here means everyday collaboration between unequal partners. Cooperation is sometimes feared and is suspected of being another word for loss of control and command. Cooperation is denounced as something for wimps who shy away from a good fight and a vigorous use of elbows. Accordingly, cooperation is misunderstood as nice but harmless. This is wrong, though.

Cooperation has to do with mutuality as well as confrontation. Every consensus that I know of builds on prior conflict, and the same is true of cooperation: mutual sharing of benefits is part of a healthy understanding of competitiveness. In general, cooperative behavior in nature is at the core of every species' survival strategy. Why should it be any different in the species Homo sapiens? The practical experience, at least in the case of cooperation strategies for sustainable development, is not so simple. The sustainable development community, with good intentions, tends to overload processes and work plans. Collaborative or cooperative efforts therefore may be subject to a fairly ambitious set of criteria: Participants can be promised equal rights regardless of their formal status. Transparency rules are set to ensure open and fair dialogue with full access to information for all. At the same time, full ownership and shared responsibility are sought. These are just examples. Each individual aspect may make a lot of sense, but they do not work in a comprehensive starter kit for action. Any hundred percent list would ignore the practical, complex trade-offs and subtle interrelationships among these aspects.

The selection of cooperation partners raises another question that is of utmost importance. It is the question of who is eligible for participation and who is not. Of course, the selection should be transparent in order to create a meaningful and reliable atmosphere. This alone is not easy to achieve. But furthermore, any selection process can unwillingly create a list of people/organisations who feel excluded, even though their participation was never directly rejected. Indirect and "perceived" impacts are often a very sensitive issue. They are often completely overlooked or potential consequences are not closely monitored and taken into account. Ultimately, unrealistic expectations do the opposite of what they were created to do.

- Conclusions

Failures are caused by the idealisation of concepts and their distance from the experience of management. Idealistic concepts lead to bureaucracy and paperwork inflation. Excessive bureaucracy is a sign for bad leadership or/and a lack of trust. Feasibility studies, concept papers, screening guidelines, safeguards, impact assessments, control hubs, check lists, monitoring, performance evaluations, engagement review tracking sheets, due diligence sheets, reporting templates, etc. consume time and resources, not to mention the (in)appropriateness of goals and means. Again, each one of them can be decisive and effective, but a compilation of all rules ends up in only expressing a sense of uneasiness and mistrust that dominates the whole process and creates even more mistrust and bad feelings.

Verification, i.e. audits by independent and high-ranking bodies "from outside" could help, but has not been developed. What role the audit offices and Supreme Auditing Institutions can play here remains to be seen. Another option is to introduce sustainability indicators into government's financial budgets, for example as carbon ton reduced by Euro spent on subsidies, or value created by green handprint built up by green infrastructure. However, comprehensive sustainability output balancing does not even exist in large companies.

Beyond the old turf, governance for sustainable development should place the highest demands on the individual competencies and skills of the staff as well as on the knowledge and insights of the central players in politics, business and society. This are more important than formalised process steps of the evergreen governance tools. These requirements are not only directed at the state and public administration and at companies (corporate environmental and social governance). Rather, they also apply to organised civil society (non-governmental, foundations, charities, academia).

Command and control, alignment, market-based self-regulation, financial subsidies and incentives will continue to find meaningful applications. But it comes down to navigating within these options in new ways. Meuleman (2013, 2018), a member of the United Nations Expert Panel on Governance of Public Institutions for Sustainable Development, introduces this as a new form of meta-governance that has a technical as well as a cultural character. New technical opportunities arise primarily in the area of digitalisation in connect to sustainability. New cultural possibilities are to be found above all in a cultural solution, which generates and supports hope, encourages personal responsibility, takes walk-the-talk seriously, and focuses on omission/non-action as well as active action. Above all, it is a matter of facing the run-away problems thoughtfully and without ritualistic know-it-all.

7.4 TINA lies, and what to do about this

We often hear that something has no alternative. For example, it is claimed that solutions for climate neutrality and sustainable development can "only" be global and that nation-state approaches are generally doomed to failure. Ambitious goals can "only" be properly implemented by institutions because only they have legitimacy and only the public sphere can drive coherence vertically and horizontally. Thus, comprehensive mandatory regulation is supposedly without alternative. Frequent reference is also made to the unique and crucial role played by cities and municipalities, because only the local level is in touch with people's lives and work. The private sector persists in portraying its role as the one solution provider. Tackling climate change, it argues, "only works with, not against" industry. The same formulation is found when replacing industry with consumers, indigenous peoples, or other social groups. The civil society sector insists that there will be no sensible solution without participation playing a dominant role and politics ensuring full accountability of entrepreneurs. Last but not least, cultural practitioners and the arts demand that culture be added as a fourth dimension to the three dimensions of economy, ecology and society, because only "with" culture and the arts can the world change for the better.

Leaders sometimes assert that "there is no alternative" (the TINA principle) when they want others to follow them without questioning their position and arguments. Those who are not in a leadership position tend to pick up the TINA assertion, but from a different background. For them, TINA is a defence of their laziness and represents an argument that prevents a change in their attitudes and usual behavior. In this sense, the TINA assertion is used as pure power language.

Correspondingly, TINA is usually directed at an audience that is uniform. It is directed at the crowd of the already converted and confirm already established opinions. Rarely is there a cross-reference to contrasting "no-alternative" assertions. TINA tends to subdue an informed debate, while deliberation, the opposite of TINA, opens up debate.

TINA stands for unimaginativeness. It stigmatises thinking and fixes the future on what seems to be inevitable. Where the talk of alternative-less-ness prevails, politics becomes an anaemic category. Bloodlessness and distance from the reality of life is the worst diagnosis that can be made about politics. TINA lies.

Encouraging political practise for sustainable development begins with challenging, disputing and contesting the TINA principle. And, of course, there are always alternatives. There are only very few exceptions (like this one: for humankind there is no alternative to planet Earth). What needs to be discussed are the preconditions, side effects and ancillaries of alternatives, and the personal perspective needed in order to discern the good and the evil, the normal way and the new normal, and so on. Thinking out-of-the-box and being able (and encouraged) to go off-script are key skills for activists.

The key to the future of sustainable development policy lies in recognising different lines of argument. Local identity, civil society participation, state responsibility, corporate responsibility or multilateralism must complement each other. This is how addressing the modern challenges of societies works successfully.

7.5 The case of the European Union

The European Green Deal set the blueprint for this transformational change. The European Commission proposes a package of missions to transform the economy, and to reach ambitious climate targets. The 27 EU Member States committed to turning the EU (meaning themselves) into climate neutrality by 2050. As an interim target they pledged to reduce emissions by at least 55 percent by 2030, compared to 1990 levels, short the fit-for-55 pledge.

The technology part, electrification, renewable energy, the Green Bauhaus building programme, circular economy, is expected to generate high employment rates. The Commission also proposes to restore Europe's forests, soils, wetlands and peatlands in order to increase carbon absorption.

This is a grand move – after years of silence and inactivity that followed a start in 2001 which was quite promising when the European Council in Gothenburg adopted the first EU Sustainable Development Strategy (see Figure 7.1). However, only a short time later this strategy's protagonists were already in a defence fight against economic growth concepts that followed neoliberal and classic concepts of economic growth and purely market driven competitiveness. The majority of administrators considered the latter, which soon became the Lisbon Strategy, as superior to the Gothenburg Strategy. There was the language regulation that the Gothenburg Strategy for sustainability and the Lisbon Strategy for growth and jobs complement each other, but they actually never did. Gothenburg withered and did not live up to expectations, not to speak of requirements. There were a few

1957	Treaty of Rome, 1992 Treaty of Maastricht, with neither reference to sustainability
2001	The Gothenburg Strategy (the EUSSD) failed. The Lisbon Strategy on (only) economics took over, and did not deliver either (2000–2010).
2004	The notion of a European Constitution (with sustainable development as principle) was rejected by several Member States.
2007	The Treaty of Lisbon abandoned the idea of a constitution.
2008–2018	European silence
2019	EU Commission announces the European Green Deal sectorised → climate → circular economy → biodiv and food systems → "Global Europe"

Figure 7.1 The case of the European Union

helpless attempts to revitalise the strategy, but they were in vain. To make matters worse, they were also very bureaucratic and thus, this strategy ended up with the image of being a puzzle of reports and desk work. Consequently, Europe as a whole did not stand out in the negotiation process for the UN Summit in Rio de Janeiro 2012. When the 2030 Agenda and its SDGs were on the table, Europe took an endless amount of time to only simply digest the new step; moreover, internal disputes began to increasingly paralyse the Union.

The European Green Deal, EGD, Europe has ended the Silence. EGD is not a one-size-fits-all solution as it is often publicly communicated. Rather, it is an overarching concept with simultaneously functioning instruments and some great lighthouse programmes and supposed game changers. With the EGD, the voice of the European Union is back on the international stage, but internally the voice remains fragile. In terms of strategic governance, the EGD draws a blank. It just does not follow up on the debate and preparation of an European SDG strategy although the conceptual framework is there (personal communication by the EU expert Inge Niestroy). The official interpretation of the Commission is that each Commissioner will ensure the delivery of the UN SDG within their policy area, with the college as a whole being responsible for the overall implementation. Apparently and knowingly, this does not work. It is another attempt at the Mikado school game, where all players feel the intrinsic motivation not to move before the others do.

In recent years, the European Council, the European Economic and Social Committee and civil society stakeholders have repeatedly and urgently called on the Commission to develop an explicit strategy for implementation of the SDGs in Europe. This has been rejected by the EC. Instead. There is no processual and institutional addressing of those issues. The EC leaves conflicts between individual goals, the lack of integration of infrastructure measures, the open questions in coordination with private financing or the timing of measures between member states to good faith. This attitude may even strengthen some conflicts of interest. This means continuing a failure. Consequently, the Commission also discontinued the multistakeholder advisory body that Vice-President Frans Timmermans used to run during the times of the Junker-Commission. It is only a guess, but perhaps the Commission feels that it is already a sufficient multistakeholder event in its own work and in the face of conflicting opinions among member states.

The EGD's financial structure comes with some new ideas. The Sustainable Europe Investment Plan 2020 aims to mobilise EUR 1 trillion by 2030. Its various funding pillars start with EUR 500 billion from the EU budget and include a new fund providing guarantees from the EU budget to support investments by the European Investment Bank and national public banks. The Fair Transitions Mechanism and Funds for Innovation and Modernization are smaller additions. For any real progress, however, fiscal approaches need to join forces with private investment, and that needs restructuring the financial industry. Currently, it does not recognise sustainability at all or it applies a rather opaque variety of self-invented standards for sustainability. EFRAG is supposed to present European sustainability reporting standards that would change the accounting standards of IFRS. The EU Taxonomy Regulation is supposed to boost climate change mitigation and adaptation while avoiding significant harm to other environmental objectives. The Commission admits investments in nuclear energy as climate-neutral and hazard-free. This undermines the credibility of the taxonomy. Even basic risk assessments will keep private equity away from open-checkbook situations, such as those that regularly occur in the construction of nuclear power plants.

Politically, the EGD suggests a different reading of the paradigm of an *ever-closer union*. This paradigm was coined by European leaders over decades and it expresses their belief that every inner-European issue can, finally, be solved by both bringing in new member states and engaging members in politics that allocates ever more competencies with the European institutions, parliament, commissions, and council of member state leaders. With the British exit and member states that are pushing ahead with the dismantling of the rule of law, the internal cohesion of the Union is under ever new strain. Countries that openly consider themselves as illiberal democracies are questioning Brussels level of democracy, and the level of sustainability as well.

The ever-closer notion seems currently outplaced. For the time being, an ever-closer cooperation (between members that are willing to do so) seems the maximum effort. The paradigm switch comes not by surprise. The notion of an *ever-closer cooperation* expresses a reinvigorated role of the national state. That is certainly a special brand of cooperation that does not make the implementation of the Green Deal easier, neither that of the UN SDGs. The point is that even the cooperation mode requires institutions that are capable of action and willing to assess their strengths and weaknesses properly. They must be able to deliver what they promise.

Interestingly, the official Europe is not good at realising what happens in its direct neighbourhood, namely in Africa and the Arab States. According to Zumbrägel (2020) Arab states are eager to announce ambitious goals, but frequently fall short of their commitments, with the exception of the United Arab Emirates, which he sees as the region's poster child with its long-term and inherent environmental vision. The oil-rich Gulf monarchies rank middle-average in the SDGs performance, on a similar level with the Maghreb states, Lebanon, Egypt and even war-torn Syria. For all of them, the climate threat is looming, and the memory of the Arab Spring will soon be washed away by the Arab climate emergency. The European eyewash already mentioned in connection with the Desertec project remains a major problem. On the other hand, once this is recognised, cooperation can be developed in mutual responsibility.

7.6 Meet the bad and the ugly: when underachieving coincides with underchallenging

Where underachievement (failing of goals) coincides with political underchallenging (the underachiever) that creates a vicious showdown between the bad and the ugly, flattening the rational learning curve of failure. In Germany that was, and partly still is, the case with the Dieselgate, the growth debate, and the carbon offsetting of a tool in climate protection politics.

- Dieselgate

The scandal called Dieselgate is an indicator of 20 wasted years. Dieselgate began in September 2015, ironically coinciding with the adoption of Agenda 2030 and the SDGs, and as negotiations closed in on the Paris climate compromise. At that time, the US Environmental Protection Agency issued an initial notice to the German Volkswagen Group for violating US law. It later emerged that this company and other automotive companies were skirting emissions regulations by deliberately programming their engines to cheat regulatory emissions tests. This led to charges of large-scale fraud and conspiracy, as well as billions of dollars in fines and civil penalties. Beyond the economic and personal damage, Dieselgate marks an even more important and serious gap in climate strategies. The age of e-mobility could have begun 20 years earlier. The transportation revolution, with a focus on non-fossil modes of mobility, could have started 20 years ago, transforming cities, new architecture, and urban culture.

For too long, the automotive industry indulged in self-praise about the alleged environmental superiority of the diesel technology. For too long, they were let off the hook. But Dieselgate is not just a story of corporate mistakes. It also has to do with the fact that consumers willingly allowed themselves to be lied to by believing the allegedly measured environmental virtue, while at the same time being able to read the real consumption figures of their cars at the gas station. And it has to do with the political mistake that a first-class idea, the energy transition, was politically implemented only second-class and produced third-class results. Underachievement is a good reason to demand now more radical moves.

Failures result in high penalty that could have been avoided by proactive action: According to American estimates, Volkswagen paid 32 billion US dollars in penalties to customers and the US authorities alone, to which they add another 30 billion in lost market revenues. This money could have driven the transformation to electromobility years earlier and, on top of that, given Germany a network of electro-filling stations. It could have linked the turnaround to the electric car to the circular economy by recycling the materials used in the car. Penalties totalling $160 billion are being paid by major banks to US authorities in more than 100 settlements and judgments since the 2008/9 financial crisis. That's according to a 2016 report by the independent US organisation Good Jobs First, with Deutsche Bank coming in at $11.75 billion. The stock markets find nothing in penalties and the stock price doesn't even twitch. This is not good news for society and, when viewed in the light of day, not even for the paying companies.

- The GDP fatigue

If there is one single constant in the economic policies since 1945 then it is the way growth is measured by the GDP. Yet, it cannot measure whether the economy develops toward a sustainable path. The debate long goes "beyond growth" (Daly 1997; Jackson 2016, 2021; OECD 2020) and produces alternatives to the GDP such as degrowth, post-growth, or green growth.

There are more than enough shortcomings of the GDP that remain unaddressed. In particular, and most significantly, the financial data of enterprises used to calculate the economic product do not adequately reflect the overall value of companies. External effects such as environmental damage and social losses, but also positive values such as reputational gains, corporate culture and innovative advantages are not taken into account. For companies, however, this is often where the risks lie, and also where future breakthroughs could be released: Human rights in the supply chain, conflict resources, assets that might end up stranded, innovative competitiveness, nature-based property rights. The greatest risk, however, is posed by those who negate the risk they themselves pose by refusing to look at those comprehensive risks. Even more problematic than the material risk is often the fact that decision-makers and business leaders are not aware of the risk they pose, and therefore cannot consciously take it into account. But today, those who fail to incorporate unpredictability (environmental/sustainability, social, geopolitical, disruption, chance, and the improbable) become unpredictable themselves.

The GDP concept did not grow along with the material problems it created. But, of course, there have been several efforts to adjust economy-alone measures indicating economic welfare, human development, the standard of living, or the gross happiness. Germany developed a satellite system to the GDP that compiles various sector-specific environmental statistics. China, remarkably, choose another way. China proposes a Green Gross Domestic Product, balancing gross ecosystem value and costs due to pollution damage and ecological degradation. Yue Pan, in his position as Vice Minister of the Ministry of Environmental Protection of China in 2005, enforced this concept and promoted it in the global dialogue "BRICS plus G" (RNE and GTZ 2005) that brought together experts and official from Brazil, Russia, India, China, South Africa and Germany to discuss the relation between sustainability and growth and how it relates to sustainable development strategies – at that time a frontrunner initiative that unfortunately was never followed up. As a result of the initiative of Pan Yue and the Chinese Academy of Environmental Planning there are now comprehensive systemic data. According to those data (Ma et al 2020a, 2020b) the cost of environmental degradation in China increased from 511.82 billion yuan in 2004 to 1,892.42 billion yuan in 2017; while its share of the Chinese GDP fell from 3.05 percent to 2.23 percent. As pointed out by the authors several evaluations for previous periods found that the annual environmental degradation cost amounted to 3–10 percent of China's GDP. Which means that those costs – if really being paid for – would almost eat up or significantly cut into the results of conventional growth metrics. The most recent initiatives do not come from academia, but by companies themselves. Companies feel challenged by the social and ecological aspects that they are not legally requested to account for. This is another signal for being underchallenged while the regular statistics and accounting schemes are underperforming. A growing number of companies wants to inform the public about environmental and social issues in addition to their financial values. And companies want to control risks and use chances that are not expressed by conventional accounting. Of course, neither all social values nor all ecological achievements can be monetarised, take the butterfly's wing beat as an example. Nevertheless, it is necessary to explore what can be known in numerical terms in order to be able to grasp everything qualitative as well. For fully accounting on equity, assets and competitiveness, quantitative and qualitative values need to be balanced.

- Offsetting carbon

Offsetting harmful emissions is highly controversial in Germany for a couple of reasons. (a) Environmental morals consider it an escape from investing into renewables. (b) The offsetting market is opaque while neither governments nor market forces ever tried to qualify and streamline it. (c) European regulatory ideals prefer the regulated compliance market of the European emission trading system. Since 2005 it is the world's first and biggest major carbon market. (d) There is no legal link between offsetting, adaptation and mitigation. (e) There is greenwashing involved in some offsetting practises.

Offsetting carbon emissions is originally a good concept. It was meant to advance fair trade and to balance of interests and opportunities between industrialised countries and the Global South. The idea was inclusive and friendly to people and nature: Certified additional carbon stocks generated by land use management in forestry and agroecology or by recycling and carbon capture and utilisation technologies should be included in economic development. This was intended to create a clean development mechanism. Instead, the reality sees malpractice, regulatory denial and bureaucracy, and material uncertainty. A non-transparent and distrustful situation defeats its original purpose and is underappreciated by regulators. This is a faulty design.

So far, only the mandatory compliance market in emission allowances is regulated as part of EU emissions trading. It affects power generators, industry and parts of the aviation sector. They emit 40 percent of Europe's total emissions. The EU reduces the number of such rights each year to provide an incentive to cut emissions. Currently, 12,000 companies hold such CO2 rights for their emissions.

Companies can sell CO2 rights they no longer need. The right to emit CO2 will cost more than 50 euros per metric ton in 2021. Above this level, it is expected that there will be increasing investment in climate-friendly alternatives. The European Commission wants to set up separate trading for buildings and transport, because here the costs of reducing emissions are much higher on a sector-specific basis compared with industry.

The market for voluntary offsets is unregulated, non-transparent and unreliable. It is also weakly developed and lacks structures and overarching safeguards. Understandably, stakeholders and ambitious companies are suspicious. Greenwashing is in the air. The bad experience with the former Kyoto Clean Development Mechanism leads to little confidence in further attempts. For example, there is a lack of accounting rules on how companies should include natural CO2 sinks in their financial and carbon balances, which could provide a level playing field. Offsets are sometimes used as an excuse to save on direct emissions avoidance costs. Companies that strictly adhere to the requirement to first do everything technically possible to avoid greenhouse gases in order to only compensate for the remaining "rest" face problems. After all, who was going to ensure that these hierarchical levels were adhered to, and who decides whether the "remainder" is really a remainder? How to regulate the state-of-the-art of the best available techniques in a binding and reliable way is shown by the technical environmental protection for industrial plants: The European guidelines on "best available techniques reference documents" work. For the most important industrial sectors, they describe the state of, e.g. energy efficiency, industrial cooling or dust levels. They do not cover the compensation of climate performance and carbon farming.

Today, the voluntary carbon market trades almost exclusively credits generated by projects such as reforestation, dissemination of solar ovens, or education. The quality of credits differs substantively depending on the specific project, and so does the price per carbon ton. They range from ridiculously low prices to prices equivalent to those in the compliance market. They pose reputational risks when additionality and the baseline for natural carbon stocks are unclear. Credits that can demonstrate additionality, conservatively calculate their impact on nature, and eliminate double counting lead to higher prices and provide greater confidence in the measure.

The market is disrupted by old offsets that the nascent market has produced without any realistic benchmarks. Market participants fail to remove them from the market and instead continue to generate cheap projects that are feared to counteract the incentives of the regulated market to reduce and mitigate carbon emissions. The "anyway" argument, brought forward by critics, is strong: renewable energy projects on the offset market are rarely additional and probably would have been built anyway; forests receive carbon credits but are legally protected for biodiversity anyway, and the money for the credits is not needed to save them from deforestation. Trees used to generate offsets are falling victim to major wildfires, or to flooding. Given ecologic site conditions, an inadequate selection of tree species may be planted and will not really achieve carbon deposits.

Numerous not-for-profit organisations are generating and certifying carbon credits. Standards and benchmarks exist and have improved over the years, but it is doubtful that market participants themselves pay for the semi-regulatory development of benchmarks. Accuracy can be doubted. Compliance is weak as no third party monitors it.

The German debate, rather than stressing those market failures, considers offsetting a counterproductive and worthless exercise and calls it "indulgence trading". Similarly, the high influential international Science Based Targets initiative rules out counting offsets toward corporate net-zero targets. That may look good, but is delusory and squandering progressive solutions. A modern metagovernance-type regulatory approach to mature the market is needed.

The Paris Climate Accords omitted any rule-making on how to allocate and share climate benefits and how to engage third parties. The Glasgow climate conference 2021 has produced a so-called Rule Book on article 6 of the Paris Agreement endorsing nature-based solution and the cross-border benefit sharing. The German Environment Agency published a policy paper (Reise et al. 2022) deemed to guide further arrangements that are needed on the way to a non-greenwashed climate neutrality. Adaptation of infrastructures and habitats to climate change and protection and restoration of biodiversity in degraded areas must be considered on an equal footing with mitigation (reduction of carbon emissions). For any industrialised and densely populated country, climate compensation mechanisms are indispensable because there are economic productions and forms of settlement whose possibilities to cut carbon and replace fossil infrastructures are limited for ultimately physical reasons. Compensation, if properly managed, is also a proven means of development assistance. It can channel money to the Global South, increase competencies and add to mutual understanding of common goods. At the end, German environmental politics will be forced to put an end its policy to factually rule out any compensation schemes and offsets markets, and instead to actively challenge the market.

These three areas – Dieselgate, GDP accounting, offsets – show how damaging the link between underachieving and remaining unchallenged is.

7.7 Shades of learning from failures

Nothing is simply just black and white. The shown failures (and a couple of more, mentioned in some of the other chapters of this book) prompt lessons learned.

First of all, the German case of a sustainability strategy with centralised responsibility and a gatekeeper modality is itself an example of organisational learning, because everything else in the form of siloed, departmental approaches has failed before.

New governance features are being tested that go well beyond silo alignment approaches, dialogic multistakeholder engagement is replacing command-and-control policies and this is leading to reasonable and stable compromises for phasing in agricultural biodiversity, phasing out coal, and other issues. Regional networks for sustainable development find acceptance and a longer-term finance background. European institutions are committed to the Green Deal.

In the German case of sustainability politics, the concept of a sustainable and climate-neutral German administration gains ground and demonstrates a high level of motivation on the part of those involved. The number of measures and thematic breadth is constantly increasing. The actions do not wait for state regulation and political guidelines. People are taking sustainability into their own hands as an issue. Today, the need for sustainability and climate protection is rarely questioned in principle. According to opinion polls, the vast majority is no longer concerned with the "whether" of sustainability, but with the "how".

The German government speaks of a lack of impact of its strategy. A "business as usual" approach to sustainability policy would not achieve the goals set for sustainable development. Governance reforms are well under discussion, and there is no shortage of proposals, including a transformation minister and a more effective way of scaling up existing structures. A more effective policy on sustainability needs the function of a minister of state in a central position. For other policy areas, this has long been the norm in Germany (Europe, digital, culture, federalism). The German constitution should incorporate a requirement for sustainability and a reference to the principles of sustainable development. Off-track policies must be sanctionable.

Desertec in its new format is another sign that good lessons can be learned. German conglomerate ThyssenKrupp recently joined Saudi Arabia's Acwa Power and Areva as a lead partner in Desertec. ThyssenKrupp, still producing steel from coal, is a pioneer in the hydrogen economy (green steel) and is also producing and scaling the technical equipment to produce hydrogen. So there is a big learning curve. The first approach, formerly known as the Desertec Industrial Initiative, tried to simply generate green power from the desert region and transport it to Europe, which didn't work and German industry pulled out. The lesson learned is to produce electricity not from the desert, but in the desert and for the desert, and is now called Dii Desert Energy. The goal is to convert energy systems in the desert areas of the Arab world from the current fossil fuel energy base to a zero-emission supply. No longer drilling for oil, but delivering green electrons and green molecules is the order of the day. Digitisation of demand-side management and the hydrogen economy open new doors to a more flexible and climate-friendly use of electrical energy by combining renewable technologies and a change in cultural patterns.

Once again, stupid mistakes are best forgotten quickly. Smart failures are a valuable asset. Failure mining should be made an element of any sustainability strategy, with special skills, training, and coaching. These assets don't belong in the display case of a museum where they are only admired. Failures are too precious to just embarrass and ask for apologies. An enlightened society that is serious about moving toward sustainability puts them on the collective table and changes the seating arrangement.

References and recommended reading

Avery, Gayle and Harald Bergsteiner (2016) *Sufficiency Thinking: Thailand's Gift to an Unsustainable World*. London: Georg Allen & Unwin.

Bertelsmann Stiftung (ed.) (2013) *Winning Strategies for a Sustainable Future*. Reinhard Mohn Prize 2013, Publisher Bertelsmann Foundation: Gütersloh (responsible: Armando García Schmidt, Bertelsmann Foundation), www.bertelsmann-stiftung.de/fileadmin/files/Projekte/31_Nachhaltigkeitsstrategien/Case-Study-Bhutan_Reinhard-Mohn-Prize-2013_20131016.pdf

CEPEI (2021) *Main Outcomes of the Regional Forums on Sustainable Development 2021*, https://cepei.org/en/novedad/regional-sustainable-development-forums/

Clark, Hellen et al (2018) *The 2018 Peer Review on the German Sustainability Strategy*. Self-Published by the German Council for Sustaiable Development. www.nachhaltigkeitsrat.de/wp-content/uploads/2018/05/2018_Peer_Review_of_German_Sustainability_Strategy_BITV.pdf (authored by Helen Clark (New Zealand, former UN DP Administrator), Adolfo Ayuso-Audry (Mexico, government), Karl Falkenberg (Brussels), Julie Gelfand (Canada, Supreme Auditing Body), Virginie Helias (Switzerland, Procter & Gamble), Lailai Li (China, World Resources Institute), Namhla Mniki-Mangaliso (South Africa, African Monitor), Joost Oorthuizen (The Netherlands, Sustainable Trade Initiative), Teresa Ribera (Spain/France, IDDRI), Jan-Gustav Strandenaes (Norway, UN Major Groups expert, Pure Consultancy), Farooq Ullah (United Kingdom, UK Stakeholders for Sustainable Development)

Daly, Hermann E. (1997) *Beyond Growth: The Economics of Sustainable Development*. Boston: Beacon Press.

de Vries, Mischiel (2019) Serving the Science, Society, Policy Interface for a Better World: An Overview of Advisory Councils on the Environment and for Sustainable Development in Europe. *Living Document 1.0*, https://eeac.eu/wp-content/uploads/2019/11/Advisory-Councils-on-the-Environment-and-for-Sustainable-Development-II.pdf

European Advisory Councils on the Environment and for Sustainable Development, https://eeac.eu

European Sustainable Development Network (ESDN), www.esdn.eu/

German Council for Sustainable Development (2010 and following years) *The German Sustainability Code*, www.nachhaltigkeitsrat.de/en/projects/the-sustainability-code/

German Council for Sustainable Development (2017) *15 Years of the RNE*. A Review of Insights with a Foreword by the Council's Secretary-General, Prof. Dr Günther Bachmann, www.nachhaltigkeitsrat.de/wp-content/uploads/2017/11/20170911_15_years_RNE_reflection_exzerpt_for_Peer_Review.pdf

Gottenhuber, S. and E. Mulholland (2019) *The Implementation of the 2030 Agenda and SDGs at the National Level in Europe – Taking Stock of Governance Mechanisms*. ESDN Quarterly Report 54. Vienna: ESDN Office, December.

Guterres, Antonio (UN SG) (2021) *Our Common Agenda*. Report of the Secretary-General Published by the United Nations, New York: United Nations, www.un.org/en/content/common-agenda-report/assets/pdf/Common_Agenda_Report_English.pdf

Hodgson, Camilla and Billy Nauman (2021) Carbon Offsets: A Licence to Pollute or a Path to Net Zero Emissions? A Task Force Launched by Mark Carney Is Seeking to Bring Order to a Market Criticised for Its Quality and Lack of Regulation. *Financial Times*, September 1, https://financialpost.com/financial-times/carbon-offsets-a-licence-to-pollute-or-a-path-to-net-zero-emissions

Jackson, Tim (2016) *Prosperity Without Growth: Foundations for the Economy of Tomorrow*, 2nd edition. London: Routledge Publishers.

Jackson, Tim (2021) *Post Growth: Life After Capitalism*. London: Wiley Publishers.

Ma, Guoxia et al (2020a) Framework Construction and Application of China's Gross Economic-Ecological Product Accounting. *Journal of Environmental Management* 264, https://doi.org/10.1016/j.jenvman.2019.109852

Ma, Guoxia et al (2020b) The Valuation of China's Environmental Degradation from 2004 to 2017. *Environmental Science and Ecotechnology (Journal)* 1, https://doi.org/10.1016/j.ese.2020.100016

Meuleman, Louis (ed.) (2013) *Transgovernance: Advancing Sustainability Governance*. Heidelberg: Springer Verlag, https://link.springer.com/content/pdf/10.1007%2F978-3-642-28009-2.pdf

Meuleman, Louis (2018) *Metagovernance and Sustainability: A Framework for Implementing the SDGs*. London: Routledge.

Niestroy, Inge (2005) *Sustaining Sustainability – a Benchmark Study on National Strategies Towards Sustainable Development and the Impact of Councils in Nine EU Member States*. EEAC Series, Background Study No. 2. Lemma, Utrecht, 308, www.ps4sd.eu/publ/2005-sustaining-sustainability/

Niestroy, Inge, E. Hege, E. Dirth and R. Zondervan (2019) *Europe's Approach to Implementing the Sustainable Development Goals: Good Practices and the Way Forward.* Brussels: European Parliament, Policy Department for External Relations, PE 603.473, www.ps4sd.eu/wp-content/uploads/2019/02/2019-Niestroy-et-al-Europes-approach-to-implementing-the-SDGs.pdf

Niestroy, Inge and Louis Meuleman (2016) *Teaching Silos to Dance: A Condition to Implement the SDGs, IISD SDG Knowledge Hub*, https://sdg.iisd.org/commentary/guest-articles/teaching-silos-to-dance-a-condition-to-implement-the-sdgs/

OECD (2019) *Governance as an SDG Accelerator: Country Experiences and Tools*. Paris: OECD Publishing, https://doi.org/10.1787/0666b085-en

OECD (2020) *Beyond Growth: Towards a New Economic Approach, New Approaches to Economic Challenges*. Paris: OECD Publishing, https://doi.org/10.1787/33a25ba3-en

Partners for Review, www.partners-for-review.de/

Reise, Judith et al (2022) Nature-based solutions and global climate protection. Assessment of their global mitigation potential and recommendations for international climate policy, UBA Policy Paper 01/2022, commissioning Öko-Institut Berlin and Ecologic Institute, https://www.umweltbundesamt.de/sites/default/files/medien/1410/publikationen/2022-01-03_climate-change_01-2022_potential_nbs_policy_paper_final.pdf

RNE and GTZ (German Council for Sustainable Development and GTZ) (2005) BRICS+G Sustainability and Growth. Brazil, Russia, India, China, South Africa and Germany in Dialogue on Sustainability Strategies. A Conference Report by German Council for Sustainable Development (RNE) and Deutsche Gesellschaft für Technische Zusammenarbeit (GTZ, today GIZ), www.nachhaltigkeitsrat.de/wp-content/uploads/migration/documents/bricsplusg_booklet_09.pdf

Statistisches Bundesamt, Umweltökonomische Gesamtrechnung, www.destatis.de/DE/Themen/Gesellschaft-Umwelt/Umwelt/UGR/_inhalt.html

Szabó, Marcel (ed.) (2021) *Intergenerational Justice in Sustainable Development Treaty Implementation: Advancing Future Generations Rights Through National Institutions*, Cambridge: Cambridge University Press.

UN CEPA, UN Committee of Experts on Public Administration, https://publicadministration.un.org/en/CEPA

UN OSD, UN Office for Sustainable Development, https://unosd.un.org/

Zumbrägel, Tobias (2020) The Looming Climate Peril Sustainable Strategies and Environmental Activism in the Middle East and North Africa. Bonn: Study, Center for Applied Research in Partnership with the Orient, CARPO Sustainability Series No. 1, https://carpo-bonn.org/

8 Moments

Facing accumulated decision needs in sustainable development in Germany

8.1 Moments

Moments, this book argues, are a social and political concept. They mark a situation in long running processes where societies are confronted with an accumulation of decisions that need to be taken.

In Earth system sciences, from ecology to public health or epidemiology and economics we know the concept of tipping point. It describes a point at which the previous development breaks. A previously straightforward and unambiguous development, it says, breaks off and changes to a new direction with altered qualities. It must be conceded that this point need not be a truly singular event, nor is the development leading to the tipping point truly straightforward and unambiguous. The Anthropocene, with its consequences for the environment and humans, is already anything but linear. The concept of tipping points uses this (inadequate) attribution only to dramatise the effect of the tipping point. It rightly does so when the changes irreversibly cross serious boundaries. Quantifying these boundaries, Johan Rockström (2009), leading a group of 28 leading academics, proposes a new Earth system framework for government and management agencies as a precondition for sustainable development.

Sociology and consumer research use the tipping point concept when a group changes its behavior by widely adopting a previously rare practice. It is obviously always a qualified minority of the population that tips the point and makes the majority adopt a new belief or a new behavior.

The proposed concept of moments does not go that far. Neither does it not finalise development nor does it attribute the consequences of decisions taken or laissez-fair politics. It stays with the holistic notion of sustainability as an open process. It just wants to emphasise the crossroad status of ripe options to be decided upon, as they accumulated over time.

The four moments described in this book can, of course, only be a selection from a long list. The criteria for selection are the greatest immediate relevance to people's everyday lives (choices, consumption) and the most pressing industrial policy challenge (innovation, regulation, circular economy). Recognising the interdependence between overconsumption, linear economy, climate and biodiversity means working on the nexus with new governance and new ideas. That

DOI: 10.4324/9781003242024-8

would be the third aspect chosen here, and of course, fourth, it all boils down to climate neutrality as a nature-based concept.

8.2 Alternative consumption

Modern consumption gobbles up resources as if they were infinitely available, thus changing the Earth in the Anthropocene in a barely reversible and, for all the gains in individual prosperity, overall negative way leading to obesity and overconsumption. Conscious consumption can change the world in a positive way. This may be hard to believe at first, especially since the main problem lies in the sheer volume of consumption itself. The German sustainability strategy measures this by indicators on energy consumption, both domestically and through supply chains abroad, carbon dioxide emissions and raw material use. The aim is to reduce the environmental impact. A hard target is not provided.

Under the heading of consumption the strategy also focuses on institutions. It counts environmental management system installed at their sites. By 2030, 5,000 sites are to be registered with the Eco-Management and Audit Scheme, EMAS. Today, there are 2,196, only slightly more than in 2005. Since its introduction, EMAS has convinced an initial number of interested companies, but then acceptance faltered. This is because there are no financially relevant effects or other, e.g. permitting facilitations associated with EMAS application. The capitalisation of a company's reputation as EMAS-certified, initially an incentive, is no longer significant as sustainability has become a much stronger differentiator in broad industries. The strategy refrains from discussing this issue, and, so, postpones long pending decisions in terms of consumer politics.

The role of public procurement (at the level of the federal government) is assessed using the purchase of environmentally friendly paper as an example, as well as the CO2 emissions of public sector motor vehicles. A target of 95 percent of environmental paper was set for 2020, but not quite achieved. In general, however, the significance of this indicator for the overall complex of "sustainable procurement" is low anyway. For car-related emissions, no quantified target is set, but rather a continuous reduction.

On a positive note, the amount of organic food is increasing. Regionality is a selling point. Companies are pursuing sustainability concepts. Around 80,000 products now bear the official organic seal, significantly more than ten years ago, not to mention relevant private labels of food companies. For 2019, the Federal Statistical Office determined the record number of 5,321 companies in Germany that use the organic seal. Among buyers, awareness of regional production and animal welfare is particularly pronounced. The Corona lockdown in spring 2020 reinforced the trend. Retailing companies in Germany, France, the UK and the USA are running sustainability projects. Some are committed to close packaging recycling loops. Germany monitors the market share of consumer products labelled with government-issued eco-labels. The target for 2030 is 34 percent. The year 2018 showed only 7.5 percent. The indicator does not monitor the product labels used by retailers although they also express the product's contribution to sustainability. A combined metric would show a higher turnout.

In textiles, too, ethical and ecological standards mark a trend that the famous fashion designer Vivienne Westwood's describes' as "buy less, choose well, make it last". The example of the fast fashion industry stands for many others. The fast fashion industry uses numerous green claims such as "sustainably sourced", "eco-innovative replacement" and "sustainably made". But the criteria are ill-defined or unsupported. According to a study of the Changing Markets Foundation (2021) more than two-thirds of all textiles are made of oil-based, synthetic fibres, and the share is expected to increase to a three-quarter by 2030. Cheap synthetics enable the business model of fast-moving fashion and dozens of collections per season that soon end up as waste. The German Sustainability Strategy emphasises the "Green Button" as the German government's textile seal. It implements the ban on hazardous textile chemicals and softeners. Natural fibres must be tested for harmful substances. Social sustainability is defined as the prohibition of child and forced labour, compliance with minimum wages and occupational health and safety. So far, however, only the sewing, cutting, dyeing and bleaching processes have been tested, not the cultivation of cotton and the weaving of the fabrics. In this respect, the seal is also incoherent because it does not integrate the existing, undisputed standards. However, there are plans to extend the Green Button.

Many consumers today say they pay attention to sustainability when buying clothes. A minority actually behave in this way. Every German buys just under 50 items of clothing every year. That is significantly less than in the USA and the UK, and still too much. The global average is at 18.7. Sales of fair fashion are on the rise in Germany, generating sales of around 170 million euros in 2020. This is more than ever before, but compared to the entire industry's sales of 8.76 billion euros, still hardly noteworthy. Nevertheless, more and more companies are striving to be successful in this market. Reputation counts. Back in 2014 – in response to the deathly Rana Plaza disaster in Bangladesh – the German government initiated a multistakeholder partnership for sustainable textiles. Participants in this partnership include 120 companies, associations, non-governmental organisations, trade unions and the German government. Together, they want to improve the global value chain of the German textile industry. Half of the 100 textile retailers with the highest sales are members of this alliance, and this half includes the most innovative companies. This underlines that sustainability is a success factor for companies in the textile industry and that this is likely to increase.

Throwing away of unused food accounts for 10 percent of global greenhouse emissions alone. According to WWF-UK and Tesco (2021) more food is being wasted than estimated – now, of all times, when the world's nations have set a 2030 target to at least halve the loss. Neither are the states on the right track, nor is the problem getting smaller: consumers, companies and farmers are destroying a total of around 40 percent of the food produced. That's more than two billion tons of good food wrested from nature with fertilisers, irrigation, gasoline, and abandonment of natural habitats.

These figures only capture what remains in the fields or ends up in the trash cans. Obesity trends are factoring in the "metabolic food waste" resulting from perverse targeted marketing of sugar, fat, and gluten, and from the mass availability of cheap foods. Overweight and obesity are the phenomena of malnutrition in industrial and developing countries. The problem has doubled in the last decade. Food in the amount of 140.7 million tons is associated with it.

Consumption in general has the one deep impact on everyone's personal beliefs and attitudes. Education for sustainability is becoming more and more important not only in schools and universities, but especially in families, and in vocational training.

Consumption strategies function as push and pull strategies. Retailers and policymakers have long given preference to the pull strategy, meaning that their offerings respond to consumer wants and needs. They have carefully refrained from any action that could be seen as prescriptive and imposing. Whether this pull idea really applies and how it comes about that consumers express their wishes is a long-standing debate. After all, in the case of the commodities cocoa, palm oil and coffee, this has led to the result that these products are now sold on the German market quite predominantly only with a sustainability certificate. This is far from being the case for other products and product groups. Retailers offer many other products first in the premium segment at higher prices and hope that over time this offer will also penetrate the mass market.

Pull and push effects, today, have to be combined in order to meet the challenge of global runaway issues. Digital technology may empower solutions. Big data and artificial intelligence make it possible and easier to value sustainability impact into products and services. Will we once earn "carbon coins", or better "decarbonisation assets" by buying climate-positive goods?

8.3 The window of circularity

In the 1970s, Germany still had 50,000 illegal landfills; today there are none. In the last 20 years, all domestic waste landfills have been closed and domestic waste is no longer dumped. Many contaminated sites have been cleaned up. Toxic waste tourism has been legally stopped. Avoiding, recycling and disposing of glass, paper, aluminium and batteries works. In a European comparison, Germany combines the lowest rate of landfilling with one of the highest shares of recycling. The EU average is far behind if one considers that the proportion of landfilling is still over 40 percent in many countries, and clearly over 80 percent in the bottom countries Cyprus, Romania, Greece and Malta. The German waste management industry's gross value added was 28 billion Euros in 2017. In 2019, 310,000 people were employed in this sector, which corresponds to the number of jobs in the energy industry. So much for the good news.

The window to circularity is open, but it is still only a window and not a passage. The moment consists in a stop and go, where the stops consist of a long list of necessary but not taken decisions. It all goes back to the 1980s, when Klaus Töpfer, then environment minister, later UN undersecretary of state and administrator of UNEP, led the German government to enact a comprehensive law with strict regulations for a circular economy. He triggered innovation and investments and made Germany the then world market leader for patents in the field of waste and environmental technology. However, in the following decade necessary follow-up decision were not taken, and so, this role faded away. Implementation stalled. The industry slowed down its dynamic development and further strong impulses failed to materialise.

As a result, the waste management sector lags far behind the top industries. Companies in electronics and measurement technology, the optical industry and car manufacturers each invest more than 14 percent of their sales in innovation, and the proportion of companies in each sector that have an innovation strategy is over 70 percent. The circular economy, on the other hand, falls completely behind. Here investment in innovation is 0.6 percent of sales; only just under 50 percent of companies state that they have an innovation strategy.

Germany is still one of the technology leaders, with a good 11 percent share of patents registered worldwide. Japan and China each have a patent share of more than 15 percent. Germany is falling behind. This shows that competition in the circular economy is increasing worldwide. German startups and some established companies certainly have good design-for-sustainability concepts, but they lack scaling. The actual use of, for instance, plastic recyclates has only increased slightly from around 12 percent (in 2017) to around 14 percent (in 2019). The plastics market is still too dependent on crude oil markets. However, the hardly increasing recyclate share in new products also has general structural reasons. This is equally evident in the market for building materials, for example. The problem is lack of political strategy and the unreasonable abandonment of instruments that would steer in the direction of higher recyclate use. The recycling quotas defined by law in Germany only measure what ends up in recycling plants and not how much recyclate is actually used in products.

Germany has no general philosophy for the future structures of a sustainable circular economy. Closing material cycles has not progressed significantly in important areas.

Germany is good at setting goals, and bad at following up. Recently, Germany has significantly increased its recycling targets: At least 90 percent of plastics must be recycled; of this, 70 percent must become secondary raw materials from 2022. By comparison, however, only 7 percent of the plastics processed in Germany today consist of recyclates. No requirements for replacing of oil-based raw material by recycling-plastics are set although industry lobbied strongly, and rightly, in favour of so-called substitution-quota. Remarkably, circular economy targets are laid down in sectoral legislation and are not part of the German sustainability strategy. This is too little to provide any impetus.

To say the least, the dynamics of climate and resource protection will fundamentally change the entire value chain in the sectors that deal with materials that could be completely circulated, such as (all) plastics, asphalt, cement and other building material, metals. In a few decades, we will probably be talking "only" about resources and no longer about "waste". Then carbon, too, will hopefully be run in closed circles. We will most probably see new problems when all industrial processes will be decarbonised, e.g. a strategic shortage of industrially used carbon. That is where new technologies come in. Even with an extensive plastics recycling in future, with numerous recycling passes, some quantities of plastics will have to be thermally or chemically recycled. New technologies are needed to substitute fossil-based plastics in a climate-neutral way, via chemical synthesis routes using regeneratively produced hydrogen and carbon.

The concept of design-for-sustainability means reduce, reuse, repair and recycle as key imperatives, with new opportunities now available that can deal with the heterogeneity of plastic waste. Chemical recycling technologies make it possible to recycle mixed or contaminated plastic waste that previously could not be recycled at all and are end-of-the-pipe resources. In countries that are still landfilling waste chemical recycling is addressing provides options. The technology will mature if applied on a significant scale. This technology could therefore fill a major gap in the recycling loop. Its full potential is not yet recognised by regulators or it is politically even downplayed as another lobbying ploy by "Big Oil". This superficial criticism overlooks the fact that new technologies are needed to keep more materials in the recycling loop. A future-focused culture of innovation and environmental stewardship must go hand in hand. The fact that this in itself also makes economic sense and promises good business fields for many companies does not mean that the circular economy will emerge and prevail on its own. This is not the case due to the competitive mechanism and consumer inertia. Competition within the economy causes single-use solutions to undermine the circular economy. Consumer inertia encourages this because throw-away options seem all to convenient. So we need a new regulatory policy for the circular economy. It must rely innovation, substitution quotas and design-to-sustainability. It must endorse and implement pull-policies (see chapter on alternative consumption). It must make the use of recycled materials mandatory. To this end, the quality of recycled materials must be uniformly labelled. Up to now, inconsistent and unclear requirements and labelling have been used. For plastic packaging, a standard from the German standards organisation (DIN 6120) has only recently become available. This is a start that is needed for all materials, just as new technological recycling options need to be brought to market for all materials.

According to estimates by the global consulting firm Accenture (Lacy, Rutqvist 2015), the potential value creation of a holistic circular economy is $4.5 trillion globally by 2030, four times the market volume achieved today. Circularity is a clear must-do, and young people's dream for professional careers, successes, breakthroughs, and a sense of community.

Especially for a high-tech country like Germany, this would actually be reason enough for a new circular economy strategy that brings together industry, innovation and environmental precaution. So far, the opportunities have not been seized, and alternatives to the governance of the circular economy have not been seen. This is a particularly cunning form of lack of imagination.

Business as usual is not only unwise, but reactionary in the face of the Anthropocene. It is growth-inhibiting and undermines the economy and society. It does not make the necessary contribution to climate neutrality. Germany can do more, technologically and in terms of cultural patterns. Circularity begins where it is not even suspected: In doing nothing. Specifically, in deliberate and conscious omission. This applies in particular to the construction of housing on greenfield sites in connection with ever larger per capita living space. Omission and reduction may seem like a crass minority opinion. But post-pandemic, there is an impetus for urban development that makes the greenfield option look old. This is represented by the home office, e-mobility and online economy, as well as adaptation to climate change and emerging water scarcity, and the drastic restructuring of building resources. These are requests to engineering design. The aesthetic and functional integration of urban agroecology (including new forms of protein and vegetable production) and so-called net emission reduction techniques is also becoming their theme. Concrete must be recycled to a high quality; only in this way will it become climate clay. Sheetrock must not be allowed to disappear in construction waste, but must be recycled – especially since the gypsum from flue gas desulphurisation will soon be a thing of the fossil past anyway.

The notion of a "circular Bauhaus" (German for "building house") connects circularity to the recent EU initiative "The New European Bauhaus". With this interdisciplinary initiative, the EU wants to bring the Green Deal to the building industries and to urbanism. It convenes a space of encounter of art, culture, social inclusion, science and technology. Founded by architect Walter Gropius the historic art school Bauhaus was operational from 1919 to 1933, and influenced the arts, design, architecture and industrial production, while housing and construction in general account for no less than one-third of total carbon pollution.

Copenhagen, Paris and Barcelona are well known for sustainable urbanism. Germany benefits from these advanced examples, and also has an impressive line-up of pioneers: winner cities of the German Sustainability Award. Emission-free mobility, socially responsible building clusters, data autonomy in smart city concepts, water management by urban sponge concepts, urban biodiversity, high-tech building solutions with timber, aesthetics of sustainable buildings, recycling of grey energy (grey = energy that is embedded in concrete and stones and need to be saved when an old building needs to be replaced) – the German Green building certification system, a private sector and multi stakeholder initiative, implements all those facets of sustainable building. Worldwide, more than a hundred building certifications systems exist, while the most popular ones are the British and US green building schemes plus the German DGNB.

Circularity is an Anthropocenic request to consumption and production patterns and to the art of construction and to collectively advance governance of regulation, push and pull factors and entrepreneurial pioneering spirit. A circular car, a fully recyclable car, may be somewhere on the horizon, but it is still a long way from being practically available. Is circular carbon too much asked for? Can circularity in the building sector become reality more quickly? Benefits are tangible, competencies and skills are there. For sustainable development politics, the moment is there.

8.4 The Cinderella moment in climate policies

Everyone is talking about climate neutrality these days. In Germany, a lot of businesses and institutions proclaim the achievement of climate neutrality as a goal for the next few years, if not claiming to be climate neutral already.

Undoubtedly, any climate policy must follow a hierarchy of tools. The first focus must lie on mitigation, with an emphasis on phasing out coal combustion and all other GHG sources. From this perspective, adaptation is a lesser option and seems to warrant only a second, much lower priority. Adding to this downgrade, adaptation has long appeared as a mere rejection of real change, and there have been fears that adaptation is merely a cover for those who wish to resign themselves to climate change rather than prevent it. These objections are no longer justified, if they ever were. This has become obvious in light of recent signs of climate change – forest megafires, heat waves, flooding, methane emissions, and other scientifically measurable impacts. Climate change is "here to stay", and at best it is a matter of preventing even more severe impacts while coming to terms with the consequences that are already occurred. For this reason alone, climate policy must reorder its priorities. This reordering also concerns another issue that has not even been included in the priority ranks so far and would have preferred not to be dealt with at all. We are talking about offsetting emissions a land use that naturally stores carbon and renders it harmless.

To turn the tide, mitigation is imperative and adaptation is obligatory. Adaptation must no longer be the ugly sister of mitigation. The prevailing view is still erroneously that climate change negotiators are waging a heroic battle against global warming, while the adaptation community is more or less assigned the role of Cinderella. The prevailing view is still erroneously overwhelmingly that climate change negotiators are waging a heroic battle against global warming, while the adaptation community is more or less assigned the role of Cinderella. in the process, it comes down to Cinderella. Even if harmful emissions were halted in line with ambitious targets, the world will have to adapt to the effects of the fossil fuel era for a long time to come. Carbon dioxide molecules remain stable in the atmosphere for a thousand years, accumulating over time. Thus, the end of fossilisation does not mean the immediate end of the processes that change the Earth' climate. Cinderella has an important job. Adaptation is life-support. Rainwater storage and flood control, cooling cities, accommodating to sea level rise, growing crop varieties, to name only some measures.

Those countries and cities that are already tackling such adaptation and using it to take transformative steps are fortunate. they are by no means all, and vulnerable regions in particular are lagging behind. The UN Environment Programme rightly points out the adaptation gap (UNEP 2021). The report also looks at nature-based solutions, which are increasingly recognised but dramatically underfinanced.

In the fairy tale, the Cinderella momentum consists in the happy coincidence of wonder and survival power. In this way, Cinderella can overcome whatever vicious punishments and handicaps befall her. In reality, we cannot hope for miracles. A Cinderella effect we need nevertheless. Adaptation and NCS must be put on a par with emission reduction measures. Nature-based offsetting damaging emissions must move from the status of an extra to the role of a major player. Even a mere supporting role would be too little. Globally, politicians have allowed valuable time to pass, while ignoring this sense of urgency, and are not yet re-arranging climate options. This time is all the more lacking because emissions are hardly being reduced worldwide and adaptation is not making any real progress.

The traditional procedures and methods of nature conservation are tried and tested, have many supporters, enjoy high recognition and acceptance, and yet: They appear powerless: Man-made, a hundred times more species are dying out than can be expected at a natural, evolutionary rate. Nature conservation is no longer up to date. If it were, there would be an International Court of Nature to judge systemic damage to nature, similar to the way the International Criminal Court in The Hague judges genocide, crimes against humanity, crimes of aggression and war crimes. Then there would have long been a climate passport for refugees whose homeland is becoming uninhabitable due to climate change, just as there was the Nansen passport for stateless people from 1922 to 1946. Those mismatches happen while the states of the world remain stuck in their traditional geopolitical power play. Protecting global nature is an untraditional, highly effective security issue that slowly gains diplomatic relevance.

Even for many conservation-minded companies, it is often difficult to bring together the technical and scientific resources required for significant biodiversity projects. Due to their long lead time such projects are often more costly than standard business-as-usual investments. Raising capital is also often a problem because risk-sensitive investors want bullet-proof projects and would also need to have deep knowledge of ecosystems and, for example, biosynthesis.

All too often, conservation strategies deliberately persist in stand-alone sectoral silo strategies. This stems from a history in which nature conservation has always been the weaker negotiating partner. It was often ridiculed as unrealistic or idealistic. So it was obvious to first build up a separate, undisputed area, i.e. specialised administrations, specialised expertise, specialised lawyers and rights of action, a specialised community. This path is necessary. While this path is still not complete, it has to change substantively.

In terms of specialised expertise, this now becomes more than obvious. In 1988, engaging science in the three IPCC working groups (physical basic, impacts, and mitigation) and making it a lively science platform was a very good move. An even still better idea was to provide it with a structured interface that organises the science-policy dialogue and makes mutual understanding a routine. This opened doors to the governance of climate policies, a change in attitudes toward scientific work, and an overall shift of mindsets. Without this early move, the 2015 Paris Agreement and the achievements of the 2021 Glasgow Conference would have stayed out of reach, let alone the sustainability strategies on the national level. IPCC is a role model. After more than 30 years, this model also shows its systematic weakness. Two Cinderella-type examples can be observed:

For one, the basic scientific understanding of global warming is almost complete, but despite this, work continues. The IPCC Working Group Report on the Physical Basis, published in August 2021, is a statement of confirmation. It reports some scientific progress in attributing the causes of extreme weather events, as well as projections into the future. Otherwise, the science repeats what is known about the evidence for global warming with even greater certainty and with great routine. The obvious conclusion would be to break out of the 1988 established routine, discontinue WG 1, and refocus the its rich reservoir of scientific capacity on pressing questions. Thus, intensifying scientific work on adaptation would be urgently needed. The most vulnerable countries made this need clear at COP26 in Glasgow. Yet today, adaptation is being dragged along as an unloved and unclearly attached Cinderellas to mitigation. There is nothing wrong with the primary role of mitigation, except that the world has wasted too much time and done too little, so that warming is now affecting more and more people, making adaptation a matter of urgency. The research community should further the understanding of regional and local impacts of climate change, agroecologic response options, adaptation needs, and how to compensate for ecological sink and source functions. A contemporary approach to science policy would make adaptation a priority working issue.

Secondly, why did it take a quarter of a century to transfer the role model of climate science to other, not least important, fields? The IPCC's equivalent on the biodiversity side, the Intergovernmental Science-Policy Platform on Biodiversity and Ecosystem Services, IPBES, was only established in 2012. Policy advise on sustainability, in general, advise against mono-centric policy-making, and is seldom heard by scientist and activists. This is because comprehensiveness irritates the individual sectors and therefore does not automatically reap undivided approval. Protagonists may fear losing influence for their sector. That is why pretty much all policies that tackle the problems of climate change and biodiversity loss are doing so independently. Only recently, a first joint IPCC/IPBES workshop brought together climate scientist and biodiversity scientists (Pörtner et al 2021). This fact is hardly to believe. Why has it taken over 30 years and three generations of scientists for this to happen, while there is no shortage of scientifically underpinned complaints about the urgency of the matter? What about the mantra "unite behind Science", in this case?

The merging of the two scientific communities, however, represents a potentially positive moment. Obviously, policies that focus narrowly on climate may have negative impacts on nature and human ecological life support systems. Conversely, policies that focus exclusively on biodiversity may be suboptimal compared to policies that provide co-benefits. Nature-based solutions would create co-benefits for adaptation to climate change. And adaptation and mitigation would benefit from avoiding, or still better: reversing, the loss of carbon- and species-rich ecosystems on land and in the ocean. Restoring such ecosystems and measures such as agroecologic carbon farming would obviously contribute to the carbon budget still available to humankind, and would add some more time on the climate clock.

Is a future conceivable in which the contradiction between mitigation and adaptation is eliminated, in which offsetting of carbon emissions is coupled progress in reducing and replacing carbon fuels, in which for a licence-to-operate the previously mentioned co-benefits are mandatory, in which with carbon is managed in a closed cycle that no longer pollutes the atmosphere?

This future image looks a bit on the idealistic side, while, instead, political and economic refuge is sought in technological wonders of ever larger scale. Geo-engineering offers stopping global warming by shooting silver ions into the stratosphere. Direct air capture aims to remove carbon dioxide from the ambient air and – if applied on a massive scale – thus lower the concentration in the atmosphere. Large scale ocean fertilisation is being considered to sequester more carbon in the deep sea, leave alone the large techno fixes by nuclear and fusion power.

These are not at all distant conceptions. Largest-scale hydraulic engineering and concrete plans are a showcase. With mega dams and canals, Earth system technology wants to change the global water balance, and, hence, the climate regime. More than 3,500 dams of largest layout are in the planning and construction stages. Engineering and canals will divert 2,000 cubic kilometres of water per year in the future (today's total is one tenth). There is no common plan and no systemic impact assessment. Rather, each country is pushing ahead with its own installations. Quite a few are related to mining and sectoral, nation-state industries. Currently, such mega-technology is topped only by the fantasies of billionaires who believe in moving to other planets.

Politics has so far failed to specify what is meant by climate neutrality. For competitive reasons, companies and institutions currently often strive for their own balance sheet climate neutrality. There is not yet a universally accepted method for determining the carbon flux and nature-based incremental carbon reductions and how to sustain them. Criteria for measuring the effects of sinks on biodiversity, soils, and the condition of water bodies also remain to be developed and standardised.

Stakeholder and protagonists have different views on nature-based solutions (see Figure 5.2). It is commonly understood that avoidance is the top priority, followed by replacement and reduction (efficiency, sufficiency) of emissions, and offsetting and compensation by nature-based climate solutions, NCS. For all sources the independent verification of reductions is obligatory. Differences exist as to who would be in charge for this verification: the market, the respective state, or a (desirable, though still imaginary) international standard setting and control body. Credibility will stay an issue. Furthermore, natural carbon sinks that are threatened and are irrecoverable (cannot be restored if lost) pose another challenge that has to be counted in, according to a first global assessment by the non-governmental organization Conservation International (Noon, Goldstein, Ledezma 2021; see also Seddon 2020).

The leverage for solutions is the interface between market and government. The question is when the dichotomy of voluntary market and mandatory emissions trading will end. A single market is needed. Nothing else can be the ultimate goal when it comes to transforming today's dysfunctional system and expanding a solutions-driven market of handprints. The Paris Agreement is of no particular help in this case so far. It lays out a vision for cooperation that has yet to be implemented. Article 6 aims to help decarbonise the global economy more quickly and effectively by allowing countries to finance carbon reductions in other countries and count the avoided emissions (partly?) toward their own climate targets. It also provides for offsetting emissions by investing in carbon sinks. However, the Paris negotiations ended with very little clarity on how this mechanism will work. The Paris negotiators knew about the awfully complicated or fully dysfunctional Kyoto-tools such as Joint Implementation and Clean Development Mechanism. Article 6 is expected to have drawn the right conclusions. In Paris, however, it was not possible to agree on fundamental issues such as the scope and governance of Article 6. It was agreed that the operational details should be set out in a Paris rulebook. It turned out to be more difficult to agree on the rulebook than on the Paris Agreement itself.

Old and new issues continue to divide the Parties. Should we allow some double counting of emissions reductions (on the part of the country providing the financing and the country implementing the action on the ground)? How robust do metrics and verification need to be to meet minimum financial due diligence standards? Where is the line between bureaucratic safeguards and credibility of reductions? How does one deal with the problem of hot air, i.e. carryover of Kyoto-era credits? What is the best way to divide mitigation and adjustment/compensation, the latter being more difficult to define and calculate? When the ISSB was announced in the margins of COP26 in Glasgow, it became clear that it will be several years before companies can use a reliable ISSB standard, and then it remains unclear whether all countries will even join this standard. What happens in the meantime, which is so important because it is a non-delivery valley. Economic and sustainability auditors are expected to make extremely delicate decisions, and the critical public to rely on these judgments, when everyone knows that the transition is rapidly changing its terms and transition costs are accruing. Of course, auditors are paid well precisely for auditing complex issues that are difficult to see through, but that cannot be the answer.

First and foremost, this requires a changed understanding of roles and a different mindset. Decision-makers and those responsible must be able to take advantage of substantial coaching on these issues. The single market could be advanced by an independent fiduciary mechanism for climate offsets and carbon farming. It would ensure high quality of carbon neutrality services purchased and then passed on to businesses and government agencies. It would clean up the incomplete market in one fell swoop and enable it to scale. Big data management could seamlessly track emissions, sinks and the fate of carbon dioxide in the environment (tracking, verification) and serve public procurement of sustainable industrial policy. The management of carbon assets and decarbonisation could become a growth industry, provided measuring of foot- and handprint improves in a consistent way.

The biggest challenges of our time are the greatest economic opportunities of our time: net zero carbon, circular economy, nature-based food, and health care. Many important decisions to free solutions from obstacles and scale them up are pending.

8.5 Cut and COP

For German environmental policy, and even more so for the German public, the climate conferences of the signatory states are of the highest interest. They are a sounding board for commitment at the national level, and for the export industry they define the space for global competition. Unlike other issues on the multilateral agenda, climate conferences are closely followed. The public expects the government to take an active, forward-pushing stance. A failure like Copenhagen's remains in the cultural memory for a long time as a portent.

The 26th UN Climate Change Conference of the Parties, COP, in Glasgow showed ways out of the stalemate of climate diplomacy and unmet targets, but of course remained incomplete. It delivered new commitments, and promises for carbon savings. Glasgow could change future conferences. That's because the British, as hosts, took a gamble that is unusual for such diplomacy at the UN level, and it paid off. They had the heads of state and government come at the beginning of the conference rather than at the end, as had been customary in the past. This had previously only been the case in the exceptional case of Paris 2015. The high-profile appearances gave the negotiations a positive boost. The British thus shifted the political focus in favour of the major states and main emitters. Under the UN consensus principle, decisions are only made when every country agrees. This is sometimes problematic because it invites veto games. Therefore, it is right to give prestigious space to the good will of individual states or groups of states and to specifically "pick up" the major emitters with their geopolitical weight before starting negotiating the details. This initiates additional alliances and cuts in climate gases that are not binding under international law, but are of great political and material importance. The goodwill takes place – legally – outside the UN negotiations, but by using the COP as a platform, this in turn supports and strengthens the UN. Organisationally, then, this change of course is a small step, but it has a big impact. The original COP concept of a purely diplomatic event is expanded to include the modality of a platform (Which is not to say that a mass gathering of some 25,000 people in Glasgow, with all its folkloric noise and ego-shooter posturing, is approved of and is very convincing, carbon-wise).

Glasgow has brought a number of advances and impulses to protect forests, limit methane emissions, end cross-border financing of coal plants, or on the part of the financial industry. Some states have committed to a phase-out of coal-fired power generation and coal finance. Diplomats consider mentioning the phase out of coal in the final conclusions, even watered down, a good signal.

Some of the most contentious issues, including carbon budget implementation, transboundary cooperation, biodiversity, financing the trillions and compensating for loss and damage will certainly be at the heart of any further negotiations, even taking into account the partial progress made in Glasgow.

Glasgow had a second agenda: who sets the tone and determines the dynamics in climate protection? The critics from outside (too slow, too little, greenwashing) or the negotiators and the entourage of the willing? The British presidency deliberately provided a flow of success stories in order to maintain its communicative supremacy and to create a momentum. In addition to genuine breakthroughs such as the good-will declaration by China and the US, there were also more manufactured postures. They tout the perfect solution, get applause from their own camp, and make no headway on the issue. They play the usual as-if game. An example is the British presidency's declaration on the end of the internal combustion car. More than 100 signatories say that from 2040, only passenger cars with zero greenhouse gas emissions may be sold thereafter, and in the "leading markets" this should apply from 2035. However, the leading markets and largest car countries China, USA, Japan and Germany have not signed. Only those car companies have signed that only produce electric vehicles anyway or plan to do so soon. As everyone is clear about the end of the fossil car, the agreement contains a schism: More than 100 signatories say that from 2040, only passenger cars with zero greenhouse gas emissions may be sold thereafter, and in the "leading markets" this should apply from 2035. However, the leading markets and largest car countries China, USA, Japan and Germany have not signed. These are mainly car companies that only produce electric vehicles anyway or plan to do so soon. As everyone is clear about the early end of the fossil car, unnecessarily, the declaration contains a schism: It makes the undisputed goal (climate neutrality) dependent on a technology path that is well disputed. It rejects not only fossil fuels but also, perfectly, all renewable synthetic fuels. For the classical environmental groups, in line with their role, this phase-out does not go far enough. For many others, the fission mushroom cloud is an encroaching technical specification that is neither necessary nor appropriate. The result: perfection blocks progress.

In another example, too, illusionary perfection stands in the way of real progress. It is about the nature-based solutions mentioned several times in this book.

More and more governments and climate specialists are trying to link the climate and biodiversity agendas and make, for example, agroecology, mangroves and circular urban agriculture a source of carbon reduction. This was greatly obvious in Glasgow. Never before had a climate conference talked so much about biodiversity. Any artificial distinction between the two agendas, biodiversity and climate, no longer corresponds to reality. One might wonder why this link was not considered at the outset of climate diplomacy, but it should certainly be acknowledged now. After all, one thing is clear in the face of unstopped climate change: we are not in a position to choose preferable measures like certain spoons and forks from a large cutlery box. We must apply the whole cutlery box, all the time, at all stages. Cherry picking is climate luxury, and such lobbying should be given no room.

Critics, however, do just that, hypocritically objecting under the guise of perfect climate policy that nature-based solutions commodify nature. They use the term commodification as a synonym for greenwashing. This position wants the luxury of foregoing certain silverware and as justification imagines the great evil, in this case: carbon imperialism. It seeks to suppress any nature-based solution, even to deny the term itself. Instead of seeking an official definition, criteria, and safeguards for nature-based solutions, this position simply turns a blind eye to reality and tries to exclusively focus on technological fixes by renewable energies.

No regulation, the argument goes, is better than any regulation, even it is a best possible regulation.

As if in a burning glass, the Glasgow conference shows the juxtaposition of real progress in cutting carbon and ideological constructs that are obstructing solutions.

8.6 Boosting the resilient self

The Covid-19 pandemic hit the globe with a precarious economy in many cases. Germany has survived the first to third wave comparatively smoothly and has managed to keep the society and economy stable by spending taxpayers' money in unique amounts that were thought to be impossible. At the same time, this is peanuts compared to the situation in economically weaker countries and with largely informal economies. Since 2016, the number of people suffering from hunger or starvation has been rising again. There has been no progress in addressing low birth weight babies since 2012. Covid-19 is making things worse. The virus feminises risk and poverty, and 70 percent of people working in health services globally are women. They are putting themselves at greater risk, while for many of them, the incidence of domestic violence is also increasing. Economically, the impact of Covid-19 is a major setback. Some feel forced to reduce the already limited funds for climate protection, biodiversity and sustainability. It increases the criminal energy of corruption and poaching. COVID-19 returns the poorest countries, as if with the stroke of a pen, to a low level believed to be past.

On the other hand, the same challenges also lead to positive effects. Vaccines were developed within a very short time and by innovative new methods, the BioNTech leaders Uğur Şahin and Özlem Türeci have been honoured with the German Sustainability Award 2020. In German daily life Big Data innovations experienced a boost. Citizens cooperated during the lockdowns. Only a minority allowed themselves to be talked into awkward conspiracy theories and refuse vaccination. Germany and Europe avoided the economic collapse with great determination. Germany is actively seeking to get the international community to do significantly more to combat the polypandemic, and is particularly active in the One Health initiative (BMZ 2021). The European Union uses the pandemic low for a European Green Deal. Investments in the future in environmental protection, key technologies, innovations and digitisation are particularly worthy of mention here.

The resilience concept took its start in health research when it was noticed that under the same conditions some people stay healthy while others get sick. It is associated with Aaron Antonovsky, who was born in the United States in 1923. In 1960, he immigrated to Israel and worked at the Israeli Institute for Applied Social Research and the Institute of Medical Sociology at the Hebrew University of Jerusalem. Antonovsky found that 29 percent of women who survived Nazi concentration camps had positive emotional health. This means that 29 percent of the survivors were not emotionally affected by the physical and psychological terror and stress. This finding prompted Antonovsky to think about resilience and ask questions like: Why do some people remain healthy and others do not, regardless of severe difficulties and adversity? What is the origin of health? What creates health? His concept of salutogenesis is making resilience a feature for prevention. Being able to cope with difficult developments without going astray characterises resilience. Citizens must be able to participate in overcoming problems without economic-existential fear for jobs, food and housing. And their contributions must be recognised and valued. Resilient, then, is a state that organises a community effort that the majority of citizens can accept as their own.

Resilience is now a concept in many sciences and in many areas of policy. There, however, it is not a predeterminable hard physical characteristic of material, but a soft quality. It teaches us that we need to understand the differences within the system. As long as we see only the total (of societies, of average per capita carbon footprints, and so on), we may not respect the true power of change and won't get a grip on the dynamics of the momentum.

Taking resilient precautions makes sense. Nevertheless, it is just as often disputed how far the precautionary principle should go. The originally fair and productive dispute about precaution versus innovation has long since degenerated into pure ideology, which juxtaposes prohibition and renunciation versus free-riding and laissez-faire. It is therefore high time to renew both the precautionary principle and the belief in innovation. Sustainability strategies offer the suitable background to do so. Although the debate is sometimes meandering, opaque and vague, the material foundations show clear perspectives.

The precautionary principle has long been enshrined in European environmental law and in Germany. Environmental damage and burdens on human health are to be prevented before they occur. If they cannot be completely averted, they should at least be reduced to an acceptable level. The German Sustainability Strategy, in its recently updated version (Federal Government of Germany 2021, p. 91) also states that "substances may be released into the environment only in line with the precautionary principle . . . dangers and unjustifiable risks to human health and nature must be avoided". This wording is based on a statement by the Council for Sustainable Development. In addition to environment and public health, this principle is also applied in finance, infrastructure and public security.

The idea of precaution arose from the desire to prevent damage from occurring in the first place. This is where the precautionary principle gets its defensive attitude and where protagonists get the bad conscience of always being somehow too late with this approach. Methodologically, the approach is also demanding, because one has to assess the consequences of something that, at best, is not even there yet. If this is done successfully, however, it does not win approval and applause. Rather, it reduces public acceptance of the principle. The more successful precautions are, the fewer people realise that they were really threatened by something dangerous that could be averted. This paradox is well known. It also applies to individual precautions against illness and to safety issues.

Why both the paradox and the defensive character of precaution could possibly now be put to rest is based on the following observations: First, precaution is now also driven by goals of sustainability. Being mission-driven makes a big difference, as opposed to react to technological advancements. In this setup the idea of precaution is a proactive one and complements the purely reactive behavior. The proactive handprint, i.e. the sustainability solution, becomes at least as important as the footprint, i.e. the problem description. In most cases, industry is also driving its strategic investments in innovation in a targeted manner and in line with long-term goals. Likewise, foresight and impact assessments are now industry standards and no longer just academic exercises. Those links are systemic and could be used.

Second, modern basic technologies, for example in synthetic biology, carbon chemistry, biopolymers, genome editing, are used in many different areas and are relevant for both the footprint and the handprint. Basic research and application are increasingly intertwined. This is giving a boost to agile management of precaution and connect well with the overall trend toward process-oriented, co-creative methods, also using the current industrialisation of due diligence standards (industry 4.0).

Third, virtual data spaces, Big data, and artificial intelligence offer new opportunities to capture complex effects, even prospectively, and to incorporate coincidences, the improbable, and the unknown. This is also urgently needed with respect to complex trends and problems, for example for various climate geoengineering methods.

Knowledge-based institutions generally play a central role in transformation according to Jürgen Renn (2014, 2020). Their action potential is closely related to knowledge and foresight. This knowledge is never simply a constant or just consisting of zeros and ones, it rather results of the mental anticipation of chances and threats, action and inaction. Institutions are part of this cultural code, and must anticipate collective actions and their consequences.

The benchmark for action and inaction is becoming clearer. Long standing experience in impact assessments prove to be valuable and to help. It can make good use of the big difference between assuming hazards, disaster or dangerous adverse impacts to livelihoods and health, and "just" talking about an impact below that threshold, such as pollution, stress, overuse. Whether these are singular or cumulatively composed of many factors, whether they are imminent (acute) or predicted for the longer term (latent), is also important. Thus, there is potential for the precautionary principle to evolve to a procedural code: transparent, rule-based, accountable for the urgency to bring about solutions.

References and recommended reading

Accenture SE (2021) *Signals of Change: The Essential Radar That Leaders Need to See and Seize the Future, Series of Business Futures 2021*, www.accenture.com/_acnmedia/Thought-Leadership-Assets/PDF-5/Accenture-Signals-Of-Change-Business-Futures-2021-Report.pdf

Aminetzah, Daniel et al (2021) Natural Climate Solutions Are Key to Mitigation. *McKinsey*, www.mckinsey.com/business-functions/sustainability/our-insights/why-investing-in-nature-is-key-to-climate-mitigation

BMZ (2021) www.bmz.de/en/development-policy/one-health

CGRI (2020) *The Circularity Gap Report 2020: When Circularity Goes from Bad to Worse: The Power of Countries to Change the Game*, www.circularity-gap.world/2020

Changing Markets Foundation (2021) *Synthetics Anonymous: Fashion Brands' Addiction to Fossil Fuels*. Report for Eco-Age, https://eco-age.com/resources/synthetics-anonymous-changing-markets-report/

Deutscher Nachhaltigkeitskodex, www.deutscher-nachhaltigkeitskodex.de/en-gb/

Deutscher Nachhaltigkeitspreis 2008–2021, www.nachhaltigkeitspreis.de/en

Federal Government of Germany (2021) German Sustainable Development Strategy Update 2021 – Full Version, https://www.bundesregierung.de/resource/blob/992814/1942518/6a4acf041217d39bac6a81cce971381f/deutsche-nachhaltigkeitsstrategie-2021-langfassung-englisch-download-bpa-data.pdf?download=1

FAO (2011) *Global Food Losses and Food Waste: Extent, Causes and Prevention*. Rome, www.fao.org/sustainable-food-value-chains/library/details/en/c/266053/

Goldstein, Allie et al (2020) Protecting Irrecoverable Carbon in Earth's Ecosystems. *Nature Climate Change* 10, 287–295, https://doi.org/10.1038/s41558-020-0738-8

Griscom, Bronson W. et al (2017) Natural Climate Solutions, 2017. *Proceedings of the National Academy of Sciences of the United States of America*, www.pnas.org/content/114/44/11645

Intergovernmental Science-Policy Platform on Biodiversity and Ecosystem Services IPBES (2019) *Global Assessment Report on Biodiversity and Ecosystem Services*, https://ipbes.net/global-assessment

IPCC (2021) *AR6 Climate Change 2021: The Physical Science Basis*, www.ipcc.ch/report/ar6/wg1/

Lacy, Peter and Jakob Rutqvist (2015) *Waste to Wealth. The Circular Economy Advantage*. Basingstoke, UK: Palgrave Macmillan Publishers Limited

Noon, Monica L., Allie Goldstein, Juan Carlos Ledezma et al (2021) Mapping the Irrecoverable Carbon in Earth's Ecosystems. *Nature Sustainability*, https://doi.org/10.1038/s41893-021-00803-6

Pörtner, H.O. et al (2021) *IPBES-IPCC Co-Sponsored Workshop Report on Biodiversity and Climate Change; IPBES and IPCC*, https://doi.org/10.5281/zenodo.4782538; https://ipbes.net/sites/default/files/2021-06/20210609_workshop_report_embargo_3pm_CEST_10_june_0.pdf

Prognos AG (2020) *Statusbericht der deutschen Kreislaufwirtschaft 2020, im Auftrag der deutschen Verbände der Entsorgungswirtschaft*, www.bde.de/presse/statusbericht-kreislaufwirtschaft-2020/

Renn, Jürgen (2014) From the History of Science to the History of Knowledge – and Back (Neuenschwander Lecture). *Wiley, Centaurus* 57, no. 1, 35–53, https://onlinelibrary.wiley.com/doi/pdf/10.1111/1600-0498.12075

Renn, Jürgen (2020) *The Evolution of Knowledge – Rethinking Science for the Anthropocene*. Princeton: Princeton University Press.

Renn, Ortwin (2008) *Risk Governance: Coping with Uncertainty in a Complex World*. London: Earthscan.

Rockström, Johan et al (2009) Earth's Boundaries? *Nature* 461, no. 7263, 447–448, https://doi.org/10.1038/461447b

Seddon, Nathalie et al (2020) Understanding the Value and Limits of Nature-Based Solutions to Climate Change and Other Global Challenges. *Philosophical Transactions Royal Society*, https://doi.org/10.1098/rstb.2019.0120

Stuchtey, Martin, Marlene Kick et al (2021) *Making Circular Ecomomy Count: Analysis of Available Circular Economy Metrics to Develop a Practically Implementable Set of Metrics for National Progress Monitoring*. München and London: Preliminary Layout, Acatech/SYSTEMIQ, www.systemiq.earth/wp-content/uploads/2021/04/CE-Metrics-Report-1404.pdf

Stuchtay, Martin, Sophie Herrmann et al (2021) *Burning Questions: Pathways to a Circular Plastic Packaging System in Germany, by WWF Germany*, www.systemiq.earth/wp-content/uploads/2021/08/210816_WWF_Verpackungsstudie_EN_sr.pdf

Toti, Elisabetta, Carla Di Mattia and Mauro Serafini (2019) Metabolic Food Waste and Ecological Impact of Obesity in FAO World's Region. *Frontiers in Nutrition*, August 23, https://doi.org/10.3389/fnut.2019.00126

UNEP, UNEP DTU Partnership, World Adaptation Science Programme (2021) *Adaptation Gap Report 2020*, 5th edition, www.unep.org/resources/adaptation-gap-report-2020

Weigelt, Jes and Alexander Müller (lead authors) (2020) *Systemic Challenges, Systemic Responses: Innovating Adaptation to Climate Change through Agroecology*. Berlin: Self-Published by TMG Research gGmbH, https://tmg-thinktank.com/

Wiesmeier, Martin et al (2020) *CO2 Certificates for Carbon Sequestration in Soils: Methods, Management Practices and Limitations*. BonaRes Centre for Soil Research at the Helmholtz Centre for Environmental Research (UFZ), www.bonares.de

Willett, Walter, Johan Rockström et al (2019) *Food in the Anthropocene: The EAT – Lancet Commission on Healthy Diets from Sustainable Food Systems*, www.thelancet.com/action/showPdf?pii=S0140-6736%2818%2931788-4

Wolff, Michael, Katharina Kemmerich und Carsten Wundrack (2021) *Frischer Wind für mehr Nachhaltigkeit*. München: Die Rolle der Aufsichtsräte, Studie der Personalberatung Egon Zehnder und der Uni Göttingen, www.egonzehnder.com/cdn/serve/article-pdf/1618844868-28750033da3c9bf35bbc47abba9dc393.pdf

World Economic Forum (WEF) and World Business Council for Sustainable Development (WBCSD) (2021) *Natural Climate Solutions for Corporates, NCS Alliance Guiding Principles for Investing in Natural Climate Solutions*. Geneva: WEF.

World Health Organization (WHO) (2018) *Overweight and Obesity*, www.who.int/news-room/fact-sheets/detail/obesity-and-overweight

WWF-UK and Tesco UK (2021) *Driven to Waste: The Global Impact of Food Loss and Waste on Farms*, https://wwfint.awsassets.panda.org/downloads/wwf_uk__driven_to_waste___the_global_impact_of_food_loss_and_waste_on_farms.pdf

9 Democracy

No end, but rather a becoming

9.1 Fierce opponents or privileged partners: democracy and sustainability

Innovators see innovation as the source of climate solutions. Capitalists see capitalism as the one and only leverage for sustainability. Civil society activists see citizen's engagement as a precondition to sustainable pathways. Representative of cities emphasise the importance of urbanism. These are examples for a general pattern. Functional groups claim a joint cause between their work sustainable development. That is not surprising and, taken one by one, quite uninteresting. More significant is whether and how those groups approach commonalities.

What they all have in common is that they live in societies with voting right, as different and sometimes incomplete those rights may be. So, a key question is whether or not democracy is able to pave the way into a sustainable development. Clearly, democrats see democracy as a central pillar and key value for sustainable development. Others do not see it that way, or at least express doubts about it. Time and again, environmentalists sigh wistfully at the apparent ease with which authoritarian governments can decide on measures without complicated democratic processes. The underlying analysis is that democracy is part of the problem, not the solution to sustainability. Emergency regimes headed by the executive, as practiced during the hot phase of the Covid-19 pandemic, seem to prove them right.

Is democracy a natural enemy of sustainability, or are they both natural friends? Around the world, the alleged link between democracy and sustainability is being strongly questioned. The debate is characterized by enthusiasm and by strong belief in processes and people, but also by disbelief in science, distrust of institutions, and deep skepticism about liberal democracy. The latter is strategically combined with skepticism about market capitalism. Any underachieving of self-imposed goals fuels right-wing populism and nationalist thinking. Like all countries, Germany has its share of right-wing populism, but it does not play a major role in responsible German politics. In the general elections in September 2021, they lost votes and received less support than expected, at 10 percent.

Rightly, Sustainable Development Goal number 16 ventures a global and international effort to "promote peaceful and inclusive societies for sustainable

DOI: 10.4324/9781003242024-9

development, provide access to justice for all and build effective, accountable and inclusive institutions at all levels". World leaders, on the occasion of Agenda 2030, even subscribed to explicitly calling on democracy. In the Agenda's paragraph 9 they envisage a world in which every country enjoys sustained, inclusive and sustainable economic growth and "one in which democracy, good governance and the rule of law, as well as an enabling environment . . . are essential for sustainable development".

Against this backdrop, for many protagonists, democracy rightly appears as the deal maker political system through which the world's people may take legitimised decisions and may learn how to determine their fate and future. Protagonists even go further proclaiming participatory approaches to decentralised "strong" democracy and deliberative democracy. Those would be in the position to even overcome the negative effects of short-term electoral cycles and political dependency on GDP growth. Moreover, grass-roots participation gels with deep ecological awareness and some academic theories of change that are making a point in social justice, sometimes even emphasising grass-roots action over electoral politics. In this context, democracy is social exercise building on informed debate, education and a collective intelligence.

It would certainly be good if they were simply right and we just had to hope for democratic progress. But that is not the case. Democracy is neither a friend nor an enemy of sustainability. It is what it is: a reaction to oppression and defiance by the Industrial Revolution and imperial appropriation. The conceptual framework of democracy differs from that of sustainability:

- Liberal democracy reinforces individual choice and defends individual freedom – sustainability requires inter-individual cooperation and even intergenerational solidarity in the face of the challenges of global warming and other unsustainable trends.
- Democracy is a mechanism for dealing with and endlessly distributing economic wealth in a seemingly infinite world of growth. Social justice here arises from the hope for more.
- Sustainability is to become a mechanism to deal with a finite world and distribute limited supplies. Here, social justice must derive from the idea of leaving no one behind and balancing burdens and gains.
- Democracy in its predominant Western forms functions nationally and through the unspoken exclusion of economically externalised social and environmental costs. Sustainability is an outspokenly multilateral approach beyond national borders that integrates the issues of transborder supply chains and global trade benefits.
- Democracy knows how to deal with the short-term. Sustainability requires dealing with the long-term as well with short-term responses to emergencies through unsustainable trends that democracy never had to deal with.

Of course, these frameworks lead to tensions and often to misunderstandings and encroachments, but they do not lead to irresolvable contradictions, because what both have in common is process orientation and dialogue, as well as people's commitment and support. By far the most important commonality is that they both have the same and formidable enemy, namely a dysfunctional Earth. That should basically be enough to join forces and turn tensions into innovation.

The intended universality of the SDGs and the broad acceptance of the Paris Agreement attempt to square democracy and sustainability and make them privileged partners. Of course, practical parliamentary action and its ethics are not adequately spelled out which is a extraordinary field for young parliamentarians to collect and exhibit competencies. More important, though, is the moral vocabulary with which the society at large is currently comfortable but that does not do justice to the issues at hand – for which the ducking practice gives a striking example.

9.2 No ducking behind the carbon poor anymore

In the history of various footprint concepts, the ecological footprint was the first and provided the framework of thought for all subsequent derivations. In the 1990s, William E. Rees and Mathis Wackernagel (1996) began footprinting by proposing a metric representing the annual ratio of Earth's resources already exploited by humanity to the fraction the planet can possibly renew in a year. Carbon footprint reports use the unit of measure of tons of carbon dioxide equivalent emissions per year and provide averages for, for example, countries, per capita, or specific activities such as air travel, shipping, and automobiles, or products and issues such as plastic pollution of the oceans. Thanks to the historical reconstruction of industrial and consumption data, carbon footprints are available for the entire world since the mid-1800s. This makes it possible to capture the historical legacy of the Industrial Revolution.

Over the last decade, global emissions growth totalled 36.2 gigatons of CO2, 78 percent of which is emitted by just 20 countries. China is by far the largest overall emitter (more than a quarter), followed by the US, India, Russia, Japan and Germany. With its one percent share of the world's population, Germany's emissions are twice as high. Germany's per capita figure is 9 metric tons (for comparison, the Chinese population has 7 metric tons per capita, and the French per capita figure is 5 metric tons, indicating the high share of nuclear energy in the national energy supply).

Now, the most interesting thing is the per capita distribution of carbon burden within a given society. And those distribution figures are only rarely reported or spoken about in the political context of climate strategies, and climate democracy. Here is an approach to this topic.

Globally, the average per capita carbon footprint is about 5 metric tons of CO2 equivalent per year, with large differences between and among countries. Average country footprints range from 30 to 50 metric tons (large oil-producing countries with low population densities) to 15 to 20 metric tons in the case of the United States, Australia, and Canada, as well as a large number of industrialised countries with per capita values below 10 metric tons.

Average data of this type should, of course, be taken with caution, as there are countries where the internal distribution of emissions is higher than the spread between countries. The USA is a good example of this. Residents of the state of Wyoming have per capita emissions of over 100 tons annually. Even states in the Deep South of the US, such as Louisiana, come in at about 48 tons, and only a few states have a specific average similar to the calculations for the US as a whole. Other data point in the same direction, such as in India, where per capita emissions in major cities vary by a factor of ten (Ramachandra et al 2015).

Thus, per capita emissions counts clearly include more factors than just physical carbon – without being transparent. Rather, they make reference to environmental conditions (heating, cooling), levels of consumption and personal wealth, and the industrial outfits of the country or region in question (e.g. white-collar service industries versus fossil extraction industries (which are included in the footprint)). These impacts are generally not mentioned in the policy debate around climate change responsibilities, nor is data presented that would shed light on these internal distributions, such as percentile values. Both would be urgently needed, because it makes a big difference whether the average of a country is a real number that reflects the actual situation in the country, or whether the average is just a statistical phenomenon that nobody in the country feels in real life. There is no reason to believe that this will also apply to social factors such as personal income and consumer behavior.

The debate about responsibility for climate change based on average per capita data obscures reality. And more than that, it's a ploy in climate diplomacy. It allows the carbon rich to hide behind the carbon poor. It's like telling an endangered species that the total biomass average will not be affected by its complete disappearance. Or addressing the problem of obesity by implementing measure to reduce the average body mass.

A disclosure of data spreads would allow for sustainable development, or vice versa: The concept of sustainability requires transparent and public reporting on the differences between sociotopes and regions in order to build trust and clarity of perspectives and interests. Without those, the debate on responsibility will remain controversial and laden with tacit but ever more serious allegations.

9.3 One man, one vote, one carbon footprint

One man, one vote is a powerful slogan. In many countries and in many public campaigns it had triggered success for universal suffrage. The idea expressed here was particularly helpful in the era of emerging parliamentarianism and later in the phase of decolonisation and the global struggles for national sovereignty. It literally has started democracy. "One man, one vote, one carbon footprint" triggers a new debate and adds the dimensions of future and the long-term to the idea of democracy. It does not suggest coherence in the sense that there should only be one homogeneous footprint for all individuals of the world. It rather raises the question how rights relate to duties, here: the right to vote to the responsibility to protect the global common good. Today, as modern societies are challenged by unsustainable developments, the concept of knowledge democracy as suggested by the Dutch professor for public administration Roel in t' Veld (2010) becomes an ever more important addition.

Knowledge democracy addresses sharing information and spreading ideas, knowledge production and the application of this knowledge to solve problems and to advance pathways into a sustainable future. This gives approaches for democratic decision-making a new connotation and gains new relevance for societies in dealing with social justice in times of big transformations. It calls the parliaments into action. It was they who wrested from feudalism the right to assemble and the "King's right" to draw up the state budget. In many places, they still have to defend or fight for these rights today. Everywhere, however, democracy with its elected representatives is nowadays approaching another task that will develop the rights of parliaments in a similarly fundamental way.

The 2030 Agenda still takes a cautious position here. This is not surprising, since it was developed and adopted by those responsible for the executive branch, and since the legislative branch, in general, is not yet positioned to play a major role. The Agenda ask the executive branch of each country to implement the 2030 Agenda and its goals. For the national parliaments the 2030 Agenda envisages only a weak service role – supporting the implementation of the goals. It just acknowledges the obvious, alluding to "the essential role of national parliaments through their enactment of legislation and adoption of budgets and their role in ensuring accountability for the effective implementation of our commitments". The executive branch is tasking the legislative arm of democracy, while it should just be the other way around.

Studies on request of the European Commission and of the European network ESDN (Fetting (2021; Niestroy 2019; see also IPU 2016) report about a growing practice of parliamentarians in monitoring government's action. Government reports on the achievement of targets and indicators and on off-target situations are increasingly, but not sufficiently one might add, discussed in Parliaments. The German Parliament, since 2004, increasingly shows attempts to free the parliamentary role form its secondary role and to return the deliberative role to parliamentarians. The German Parliamentary Advisory Council on Sustainable Development (2020) repeatedly reiterates its assessment that Germany can do more in terms of sustainability than has been the case to date, and it calls upon the Federal Government to develop new and more ambitious national strategies. Parliamentarians produce opinions and position papers covering an impressive number of topics (all available from its website (see endnotes). They regularly invite ministers and government officials as well as experts and civil society actors, creating a centre of debate that is attracting more and more awareness. Across all political groups, this increases their fellow parliamentarians' awareness to the issue of sustainability. The body also criticises the procedural limitations of its mandate and its ability to act, respectively its commitment to expand legislative work on sustainability. On several occasions, the Parliamentary Advisory Council and the RNE jointly conducted a kind of street campaign inside the Parliament building, asking all parliamentarians about their approach to the SDGs.

9.4 The benefit of German consensual policy

Each country has its own way of shaping its political culture, mostly related to the conduct of general elections and the respective history of political culture. The specifically German political style emphasises consensus building and sets a deliberate contrast to more confrontational approaches. In sustainability politics, the search for consensus is a fundamental advantage if it does not become an end in itself and always leaves the possibility for majority decisions.

The German special brand of consensus focus is linking the consensus to a concept of normality. Possibly similar things can be observed in other countries, but without a doubt it is valid for post-World War II Germany up to the present. In West Germany (the Federal Republic of Germany), the electoral process systematically favours the formation of coalitions between different parties. Since 1945, only a few parties have carried and dominated politics for long periods. This constancy has shaped the way politics is made and how it is perceived by the opposition and the public. German reunification in 1990 did nothing to change this. Chancellor Angela Merkel and Chancellor Helmut Kohl each had a 16-year chancellorship. The Social Democratic-led chancellorships were not as long, but they also lasted 13 years (Willy Brandt, Helmut Schmidt) and seven years (Gerhard Schröder).

The common assumption that democracy, with its focus on four-year election periods, has difficulty with long-term and decade-spanning thinking is wrong. Even the conventional budgeting (conventional in the sense that external factors are largely excluded, as private capital has been doing until now) calculates future in. Financial borrowing even at levels far above the current budget and as regards time frame far above what reasonably can be overseen, is proving this point.

Abrupt policy changes and trend reversals, however, are rather alien to the orientation toward any consensus machinery. The protagonists of disruption (e.g. as regards climate policies or the debt break) are watched with suspicion and are sometimes considered "opponents of the system". Thus, fundamental alternatives and new approaches usually take a long time to gain acceptance – and are then no longer recognizable as such. They are considered "normal". For politicians seeking policy change through sustainability goals, this is a problem. To convince voters and people of the purpose of their political programme, it would be helpful if they could point to past successes where systemic changes have already led to better solutions, such as successes in the circular economy or renewable energies. These successes very much exist, but they have long been considered normal and are no reference points worth mentioning in the eyes of the mainstream. This even applies to the positive effects of German reunification on the lives of many people.

The mechanism goes like this: Majority politics emphasises the political categories of security, order, normality. It presents its policies as "there is no alternative" and rejects fundamental changes (system change). However, it observes whether any alternative idea nevertheless achieves chances of a qualified minority. Then this idea, if suitable, is adapted as its own and passed off as self-evident. This is supposed to signal normality to the citizens. Politicians connect this kind of change seldom with big picture visions. They would rather see change as a process to adapt policies to reality. Vision and long-term goal setting is not a dominant cultural feature in German politics. This sounds like bad news to those who take the 2030 Agenda seriously. But the lessons that could be learned are not entirely bad, because once the mechanisms are understood, they could be used and activated in a transformation strategy. They might even provide policy opportunities that would not otherwise be overt.

Insofar as external events are as extraordinary as, for example, flood disasters, heat waves, forest die offs or mass migration of people, these events are called "crisis". In the understanding of an exception the term crisis suggests that afterwards the state of affairs from before will return. At the very least, it raises the expectation that what is normal will not fundamentally change, even as a result of disruptions. Whether the situation before and after then really coincide is no longer relevant.

Consensus promotes a structurally conservative politics. For its invocation of the normal always applies only to the present. In this respect, the competence to deal with different options for the future and development is beside the point and is not specifically promoted. Sustainability strategies are actually alien to this attitude. Setting goals for periods that go beyond legislative periods is considered basically illusory. Long-term political planning is considered hubris. If, as in Germany, a national sustainability strategy does come about, it is seen by this attitude primarily as a technical tool for meeting current investments with long-term cycles. This attitude even indirectly influences its opponents when they adopt the same understanding of crisis and see in all problems only ever greater crises. Accordingly, they "translate" the future simply by demanding goals and what they think is desirable in the long term.

This has several consequences for politics not in favour of transformational change: (a) Real, transformative situations rarely reach the active consciousness of the public; as a result, real opportunities are not seized or not exploited. (b) Policymakers themselves do not build up specialists and high-profile experts, but "rely on science" without further ado. (c) Progressive actors defend themselves against this, in that for them change always has to come "from below" or "from outside" and has to be provocative, in other words, it has to send alarm messages. Polarisation is their quality characteristic. (d) These consequential effects tend quickly to become rituals. They lead to some time joyless, sometimes opinionated and often over-engineered debates or hollow reproaches.

The German constitutional system is pushing in the same direction. It establishes a very division of labour between the 16 states and the federal government. The states and the federal government each have their own clearly defined, self-contained legal responsibilities for which the other level does not have to be consulted. In addition, the states have the legislative right of co-decision and negotiate on this in the second chamber, the Bundesrat. They can therefore shape federal legislative procedures insofar as these concern Länder competencies. It establishes a very division of labour between the 16 states and the federal government. The states and the federal government each have their own clearly defined, self-contained legal responsibilities for which the other level does not have to be consulted. In addition, the states have the legislative right of co-decision and negotiate on this in the second chamber, the Bundesrat. They thus help to shape uniform federal legislative procedures insofar as these encroach on Länder competencies. Increasingly, legislation, because it deals with complex issues, requires close coordination between the federal government and the state governments, which usually have different political compositions. This constitutional principle promotes consensual politics.

This mechanism is a lesson learned from German history and German fascism, and rightly so. "Never Again!" is the founding idea of Germany's post-war order. It follows from it that it would be more difficult for one-party governments. Germany's proportional representation electoral system creates a multiparty system. It proportionally translates vote share into legislative seat share. This allows voters to align themselves with more diverse political parties, without the fear that their vote goes unheard because the winner takes it all. This avoids monolithic parliamentary setups and rather prefers smaller, more diverse political parties and the building of coalitions.

Thus, negotiation-style processes, as well as democratic dialogues, are intrinsic to the system. It makes top-down authoritarian governance impossible. Of course, the system also produces lazy compromises and opaque walls of feigned consensus. But for the most part, the system promotes political enlightenment and participation. Experience shows that a politically particularly stable consensus is usually based on a contentious and open conflict. This is then a healthy relationship between consensus and conflict.

Other democratic systems gear their procedures more toward contrast and the antagonism between majority and minority. Examples are many presidential governments or the electoral law of the USA. Apart from that, there are political ego currents in many countries, which are not meant here, because their neo-nationalistic and chauvinistic behavior leaves the ground of democracy.

In Germany, consensus thinking is an overriding political principle and is certainly part of the state's raison d'être. Knowledge of the historical background is alive and well. Consensus orientation is therefore hardly questioned and is part of everyday political life. That helps legitimise sustainability dialogs between politics, society, science and companies. Organised civil society is more likely to be involved in political strategies.

However, this is not automatic or a one-way street. Every step has to be hard-earned in some cases. Good arguments must also be well put forward. Sometimes there is also backsliding or the debate freezes in Germany. Then, for example, there are arguments about details or individual aspects such as the cost of gasoline, the "whether" of domestic flights, or the price of pork, without doing justice to the actual requirements of structural.

Democracy must not be reduced to voting and the works of constitutional institutions. Trade unions and the organised civil society have been instrumental in shaping the 2030 Agenda. Sustainable production, financing and due diligence along the supply chain are increasingly forcing companies to face strategic challenges. They are to review and adapt products, processes, and what is called decent work. They are to integrate sustainability into the core business case, and often this practical works the other way around: adapting the core business to the requirements of sustainability and a new understanding of competitiveness. This gives rise to conflicting goals. The interests of management, employees, employee representatives, committees, investors and NGOs differ.

The German economic system operates to a considerable extent with so-called co-determination, which is largely unknown in the Anglo-Saxon, Russian or Chinese economic world. Trade unions and employee representatives sit on supervisory boards and management boards and play an important role in addressing and implementing sustainability issues, e.g. with regard to working conditions, job creation, site changes, investments, etc. In the German context, the Metal Union and environmental NGOs are joining forces with regard to the mobility system and its necessary but not yet sufficiently implemented path to sustainability. The Chemical and Mining Workers Union focuses on supporting and coaching workers and workers' representatives to contribute to the company's sustainability agenda. Through its spokesman Michael Vassiliadis, a protagonist of sustainable development, the union participates in the political and strategic discussion on the future of the entire energy system. These are active contributions to democracy.

9.5 Power factors

It sounds strange, but democracy, although it establishes and controls rules without restriction, depends on conditions that it cannot control conclusively: that is its success and the effect with which it contributes to this success. Democracy is founded on its own success. Democracy, unlike totalitarian forms of government, requires the constant renewal of its own procedures and the continuous experience of the people that their own commitment is worthwhile. This is how consent and approval come into being. This cannot be decreed in the abstract, but must be lived. Where the latter is not the case, it threatens to wither and become emaciated.

Success is nowhere seen as concretely as in the eyes of people who are committed, reward themselves with the result, and feel part of a bigger agenda, and of a democratically shared idea for the future. This feeling makes them shine. Yet "saving the world" or some far reaching climate goals are rarely the benchmark, but often the small steps and manageable means. This may be a bicycle lane on the main road, or that less food is wasted in a canteen and that there are more organic dishes. It can be in the concrete experience of the togetherness of a city community or in the project performance of a group of students.

Twenty years of sustainability policy (since 2001) have changed more in Germany than the 20 years of discussing sustainability ideas before. Many things have changed that were considered impossible dreams at the beginning of the

2000s. At that time, sustainability was a vague concept and the term was considered almost poisoned. It was called a rubber word and should be avoided at all costs. There was no effort to communicating worth mentioning, let alone a renowned German Sustainability Award. The local agenda had no power in many places, with only a few exceptions. The mayors of cities (active in sustainability) did not yet have a common platform. Civil society organisations were not yet regularly included in the German government's sustainability strategy. There were no research programmes for sustainability. Business leaders were 99 percent convinced that sustainability burned money and was just a fashionable hobbyhorse of outsiders. Being "in charge" of sustainability was seen as destroying careers. Those who did it anyway were sure of pity and regret. The environmental movement was at war with the notion of sustainability, considered the term a Trojan horse of evil economic lobbying, and tried to sideline it.

All that is different today. Sustainability is a power factor. Every politician is asked about climate protection and sustainability. Without a presentable agenda on how Germany should face up to the demands arising from these issues, it is almost impossible to make a political career. The general election in September 2021 was influenced by the climate agenda more than anything else. Democracy has emancipated the issue. It has left the niche and reached a social breadth. The actions for education for sustainable development reach many people in schools and extracurricular educational institutions. The German Action Days on Sustainability, for which the RNE has been calling for voluntary commitment since 2012, recorded around 3,500 participating projects and initiatives in 2021. Under the heading of "sustainability culture", the Sustainability Council promotes transformative projects that drive change in everyday culture in the areas of food, mobility, fashion and clothing, etc. The sense of making a difference in small projects is widespread and strong. Being effective oneself is a basic democratic experience.

Meanwhile, mayors and campaigners are running under the banner of holistic sustainability, and they are succeeding too. Gone are the days when half-hearted greening here and social commitment there could be used for a Sunday speech on sustainability. With sustainability, actors can assert themselves in the democratic power game. In a democracy, power is ultimately what makes a difference, what is attractive and what others follow. There are similar awakenings all over the world.

There is much to suggest that the deep social power of the idea of sustainability comes not so much from the ethical imperative of knowledge, data, or United Nations resolutions alone, but from the practical successes of individuals. Of course, science, goals, and major policy decisions are needed, and we certainly need to do better in this regard. But only those who can mobilise people "from below" will ultimately come out ahead. And democracy can do that better than other forms of government that rely on fear, control or dictates. Therefore, individual skills are of great importance when it comes to collaboration, confidence-building measures and conflict resolution (see Figure 9.1). No question, there are great opportunities here to become even better.

Provided the purpose of a collaboration and its goals are clear, then the following social "soft" qualifications are critical to successful performance for those involved or partners.

1. Co-operating is a matter of trust. Trust can be given, but not owned.
2. An overstretched measure of trust is of no use.
3. Everyone is someone, meaning the chemistry factor is important.
4. Compliance culture applies to resilience and responsibility, at all levels.
5. Everyone needs a stated role that others understand.
6. As performance is always heterogeneous, the heterogenity must be laid open.
7. The flow of (almost always: asymmetric) information must follow some general rules.
8. Competences must be recognized and creativity valued.
9. Especially on "eye level" among equals leadership qualities are necessary.
10. Everything can be in flow, everything must be tested and reflected upon along the process.

Figure 9.1 The ten soft success skills

9.6 Becoming

The world has survived the hardships of the Industrial Revolution and has completed a successful civilising process. Today, it must do so again, this time the civilising effect needs to be even more fundamentally.

The first process addressed the contradiction of capital and labour and marginalised the environment. In 18th- and 19th-century Europe, capital and labour were irreconcilably opposed. The Social Question shaped the industrial revolution. It was absolutely central. The exploitation of working people was the big idea of the time. Harshly and relentlessly, ruling elites subjugated working class people to the beat and regime of machines. Exhaustion, industrial accidents and lack of hygiene claimed uncounted victims. The working class was oppressed by police and social coercion. It was led into destruction and war. More than 150 million people were victims of wars and civil wars, genocide, and famine. The memory of mass death, violence and displacement continues to shape politics in Europe to this day. The German fascism added the experience of the Evil, of terror and the Holocaust. Central for the German society after 1945 and also for many individuals are individual and collective experiences of guilt and complicity as well as with individual concealment and societal mechanisms of repression.

Historically, however, those times also saw means and ways that successfully addressed the social question. The Industrial Revolution, in the long run, has prompted a civilising process. That its results would always be just or satisfactory cannot be said. Nor did it proceed uniformly, smoothly, or peacefully. But it has produced lasting achievements, such as labour law and wage agreements, education, equal rights, a social housing policy, general hygiene and health care, a system of old-age security, pensions and health care. Institutions and legal systems have emerged to defend and negotiate claims, rights and obligations. For this civilising process to happen opposed parties (in the end) negotiated successfully and institutions stabilise and steer the negotiation with a system of rules and orders. This process is guaranteed by democracy and its rule of law. The modern type of democracy, despite all its substantive differences across the world, is a child of social unrest and revolution against the hardships caused by exploitation and false regimes.

Typically, those negotiations take place primarily in parliaments and workplaces, but also in civil society. It teaches all participants a constant balancing of give and take, virtually in the form of a trade in exchangeable values, such as labour vs. payment, taxation vs. social welfare, and so on.

In the course of the harsh and bloody industrialisation of the 19th century, this process created depth and background in theories and worldviews, left and right. Marxism, social democracy, Christian social teaching, and liberalism influenced the central doctrines for politics and government. They changed the way society thought. They addressed the questions of what justice is and what one should be free from, what adaptation and revolt mean, how to deal with illness and otherness, what demands the community can and may make on private property. The answers varied, but what remained was a deep basic understanding of social commitment, of solidarity and subsidiarity, of interdependence and commonality, which shaped the general way of thinking.

It cannot be overlooked that this has had an effect. The lives of many people have improved. Life expectancy has increased and continues to rise. Infant mortality has dropped radically. Even warlike conflicts – as violent and cruel as they are – fortunately do not have the scale today that we know from history. It is still luxury versus poverty, top versus bottom, wealth versus hunger, corruption versus fairness. In some cases, the inequalities take on bizarre proportions. The improvement of the social world occurred with a powerful lever and came at a high price. The lever was the steady growth of the economy, which, for all the privileges and wealth at the top, always created something to distribute to the masses. The price was the radical destruction of nature. So the social question must be reinterpreted. It must deal with the ability to survive on a planet changed by man. The question of the survival of the human species outweighs the question of the situation of classes. The essential contradiction today is not between capital and labour, but between them and the environment.

It must not be overlooked, however, that there are no comparably strong theories of the social today as there were then. Compared to the Industrial Revolution, today's situation is speechless, and its equally great upheavals of nature, industry and society remain strangely technical. Pope Francis's eco-encyclical *Laudato si'* remains a solitaire. Rachel Carson's *Silent Spring*, Dennis Maedows's *Limits to Growth*, the climate budget, and the Nobel Peace Prize for climate scientists with the IPCC became icons of ecological consciousness (which is good), but there they wither into mechanistic thinking that lacks a cultural praxis. There, in turn, in the practice of the many thousands of sustainability success stories – whether in small projects, marketplaces and workshops, schools, associations, startups or veritable companies – a common point of reference is missing: a constitution, so to speak, that applies to everyone and makes each individual stronger.

In the 21st century, the central question is no longer social, but socio-ecological. And now, the most remarkable change is the following. This is not any more a two-way negotiation between social partners, like in tariff autonomy, but it is still a matter of negotiated decision-making. The protagonists of a strict climate policy often say that physics cannot be negotiated with. In this view, laws of nature (physics, ecology) are a given set of rules to which humans must adhere in order to survive. Mankind, it is (in Germany very often) said, must follow the laws of nature that science has discovered, in short, people must follow science. I object. This reduces a highly complex situation. Mankind is not a subject that would act. Of course, people continue to negotiate with each other, just in different constellations, which are different than before. Natural laws are today of course just as little negotiable as at times of Galilei when it was a question whether the Earth turns around the sun or the other way around. however, today the effects of different natural laws are to be weighed with each other and effects as well as chances are differentiated regionally and with regard to the concern of people. One cannot negotiate with nature, but one must negotiate with people about nature. The sentence, you can't negotiate with nature, is misleading. And it only means that one side considers its arguments sacrosanct and wants to immunise against questioning a priori. But this is an illusion. This attitude wants to strengthen one's own arguments, in fact it weakens them.

Technically, negotiating the balance with nature requires new models and procedures. In terms of content it requires new thinking. The new thinking must be able to weight things that are not arbitrarily interchangeable. It must account for scarcities and be able to deal with imponderables and the burden of long-term adverse effects. Also, it must have an idea of how to share beneficial solutions. Finally, it must produce procedures that put legitimate interests (they still exist and are powerful) at second stage and deal with the many conflicts of interest.

Politics is lagging behind here, and strategically so. It has so far socialised its actors very successfully in the sphere where everything is exchangeable: people, parties, relationships, interests, money, power, careers, jobs and entrepreneurs. In this understanding, politics is a barter system that is good if it is organised in a comprehensible, honest and socially acceptable way. The political mainstream would be well advised to better attune their actors to these new bargaining options. Political negotiation is now becoming a difficult task that goes far beyond the schemes of climate diplomacy still in use today. The conventional political establishment is remarkably unsettled by this. Because now a new factor comes into play that differs from the previous civilising process: Properties of natural ecosystems are not exchangeable – with what should methane in the Arctic permafrost be exchangeable? Nor is nature over when functions are replaced by technology and capital – don't soils remain distinctive bodies of air, water, fungi, animals, plants, minerals and organic matter, even when their fertility is enhanced by artificial fertilisers? Nature is also much more than what man lets nature be and what man understands – soil animals, viruses and bacteria mutate and create ever new habitats. Finally, the boundaries identified by Earth system science are indeed limits – but they do not appear in the form of previously calculated linear schemes, not with full scale foresight and risk assessment. They rather appear as events, and surprisingly dynamic events at that.

Playing for time and sitting out problems is an accepted and well-functioning pattern of political practice as long as one operates in the world of barter. But the nature factor changes the concept of time. Postponing real conservation action has hardly ever really worked. Instead of the hoped-for breather, it ultimately only brings back the material problems from global warming and nature destruction all the harder, including the resulting bottlenecks and sustainability risks to the economy and health. Postponement means aggravation.

The idea of sustainability seems like a call to order. The idea reminds us that order in societies is highly dependent on the material foundations on which a society lives. It exhorts concepts for threatened orders. Between gradual adaptation, prudent small as well as large steps, and hardening into a draconian mode of defence that may degenerate into a downright lust for ruin, threatened societies have practiced many things. There is a lesson to be learned from this. But one must be able to learn by doing, to understand differentiation and to connect what exists with what is becoming.

All this shows that democracy is more than just counting votes and waiting for the majority. It goes beyond numbers and reaches deep into the natural conditionality of human life. Democracy is a driver for sustainable development. All of its value lies in the becoming. If it is true that in times of great transformations the willingness to learn socially is the key, then democracy is the best option we have to move hesitant situations into dynamic self-reflection and change into action mode. If our society is too complacent and do not act adequately, then it is not democratic enough.

References and recommended reading

Bexell, Magdalena and Kristina Jönsson (2020) Realizing the 2030 Agenda for Sustainable Development – Engaging National Parliaments? *Policy Studies* 1–19, https://doi.org/10.1080/01442872.2020.1803255

Fetting, Constanze (2021) *Parliamentary Mechanisms in the implementation of the 2030 Agenda*. ESDN Report, Vienna, August, www.esdn.eu/fileadmin/ESDN_Reports/ESDN_Report_August_2021_Parliaments_and_SDGs_Final.pdf

in 't Veld, Roel (ed) (2010) *Knowledge Democracy: Consequences for Science, Politics, and Media*. Berlin and Heidelberg: Springer, https://doi.org/10.1007/978-3-642-11381-9

IPU, Inter-Parliamentary Union (2016) *Parliaments and the Sustainable Development Goals: A Self Assessment Toolkit*, www.ipu.org

Niestroy, Inge (2019) Europe's Approach to Implementing the Sustainable Development Goals: Good Practices and the Way Forward. *European Parliament – European Directory General External Policies*, www.europarl.europa.eu

Parliamentary Advisory Council on Sustainable Development (2020) *Discussion Paper on the German Sustainable Development Strategy – 2020 Version, This and All Other Texts*, www.bundestag.de/en/committees/bodies/sustainability

Ramachandra T.V., K. Sreejith and Bharath H. Aithal (2015) GHG Footprint of Major Cities in India. *Renewable and Sustainable Energy Reviews* 44, 473–495, http://doi.org/10.1016/j.rser.2014.12.036

Seddon, Robert (2013) *We Have a Moral Vocabulary of Democratic Citizenship, and a Moral Vocabulary of Environmental Sustainability – but Can Our Ethics Encompass Both in Harmony? A Provocation, Foundation for Democracy and Sustainable Development*, www.fdsd.org/wp-content/uploads/2014/11/We-have-a-moral-vocabulary-of-democratic-citizenship.pdf

United Nations (2015) *Transforming our World: The 2030 Agenda for Sustainable Development*, A RES/70/1, https://sustainabledevelopment.un.org/content/documents/21252030%20Agenda%20for%20Sustainable%20Development%20web.pdf

Wackernagel, Mathis, Bert Beyers and Katharina Rout (2019) *Ecological Footprint: Managing Our Biocapacity Budget*. Gabriola Island, Canada: New Society Publisher.

Wackernagel, Mathis and William E. Rees (1996) *Our Ecological Footprint: Reducing Human Impact on the Earth*. Gabriola Island, Canada: New Society Publisher.

Westall, Andrea (2015) The Relationship Between Democracy and Sustainable Development: Foundation for Democracy and Sustainable Development. *Briefing*, www.fdsd.org/wp-content/uploads/2015/06/The-relationship-between-democracy-and-sustainable-development.pdf

10 Success

Navigating its elements

10.1 Sustainable solutions: made in Germany

It was not the Germans who invented the "Made in Germany" label, but Great Britain. The label tried to stop British customers from buying (supposedly inferior) German products. The idea backfired when this label became a model of success and a sign of quality. This old story is often told and never misses the message of success against all odds.

Sustainability, it can be assumed, is a similar story. The first signs point to this. What started out highly theoretical is now part of very practical considerations. What started out as an ugly duckling is now a hallmark for quality: in design, smart products, healthy living, but also innovation and high-end policies to tackle global challenges.

Sustainability is on the road to success, on both a large and small scale. This also applies to the German sustainability strategy. Without its influence, there would be no nuclear phase-out, and without a nuclear phase-out, there would be no coal phase-out in Germany. The strategy also achieved success in its individual targets and indicators: There is no other political platform that has so successfully put into practice the economical and sparing use of land as a resource. Its targets and action programmes have advanced organic farming more than all previous measures combined. The quantitative targets have created good opportunities for increasing resource productivity. The German Sustainability Strategy has encouraged experimentation with open methods of deliberation and dialogue as a problem-solving resource in the areas of agriculture and nature, industrial digitalisation, entrepreneurial leadership and sustainable management, or regionalisation of sustainability policy. The ever-continuing and reinforced strategy has promoted the credibility of the concept, encouraging pioneering companies to position themselves aggressively in the market as promoters of sustainability. Politically, the strategy has kept environmental protection and competitiveness modernisation at the top of the political agenda for two decades, convincing every government of the urgency of the issue, something no other issue or strategy has managed to do. For the political staff, the sustainability strategy is a constant learning event, and, in general, the top echelons of the German system of government have also proved open to it, with some curious exceptions. This is the

DOI: 10.4324/9781003242024-10

language of success which comes from a variety of sources, the most important being the gatekeeper concept.

Of course there are objections, and they are right, after all: Of course all this won't save the planet. Of course, unsustainable trends dominate; the world is failing in climate protection and Germany's transformation is slow and sometimes really annoying. Of course, the typical German doubting Thomases are right when they lament that what has been achieved is not perfect. But what does that mean? There is no planet B on which everything would be easier and more straightforward. Success is success, no matter how bigger and more impressive it could have been. But success is also about looking at what is possible and what has not yet been achieved, and certainly what can be concluded from failures and underachievement. Speaking of which and of the Made-in-Germany concept, this "made-in" needs a turn toward a "made-by" in order to cover the reality of globally distributed production chains. The German sustainability strategy emphasises that the doing must embrace the housekeeping, the German aid donation, as well as the globalisation of sustainable solutions engineered in Germany. The German export economy has made its successes from trading with partners in like-wise industrialised countries, or emerging economies in South East Asia. As we move away from a carbon past and toward a healthy planet future, the criteria for success will need to be adjusted. In the case of a future-proof Made-in-Germany, the greatest benefit for all will come when trading concentrates on the needs of the poor, respectively on the needs of the then-not-so-poor anymore. The logic of success can sometimes be that simple.

10.2 Why success seldom is self-evident

Sustainability strategies are supposed to be successful. Nothing else is expected of them. But success rarely happens by itself. Goals like the SDGs are set in order to achieve them successfully. Failure is not an option, it is often said, and success is only a question of finding the right means of implementation; and implementation, for its part, is only a question of political will. Success, according to the majority opinion in Germany, thus, is only a question of will. However, this is a circular argument, and it is wrong. It overlooks the fact that success is not only a matter of political will (of course it needs political backing and top-level engagement), but also of the practical skills needed to make success possible and to make it visible. After all, success is only ever success if it is so in the eyes of others.

The relationship between success and failure is an issue that is not as simple as it first appears. Successes are usually slow and small-scale; failures are fast, large and total. Success often has no clear address (success has many parents); when something fails, one asks very precisely who is to blame. Successes are easily overlooked, failures are not. Time smooths successful change into a new normal, which in turn is soon seen as failure. Success requires active effort, failure just happens. Success is presuppositional, attacking it is easy.

Success is structurally disadvantaged compared to failures. The sustainability community defines itself more by the general shortcomings in addressing the

Anthropocene than by its own successes. This is a peculiarity; hardly any other large social group defines itself in this way. Usually, any formation of informal groups relies on an attractor, mostly on success or an interesting role model. Here, it is issues, disaster and doom. This has an impact internally in this group which consequently is often overseen or not taken for serious: it is also very rare that success is accepted and appreciated as such by all those involved and affected. Successes, if there are any, quickly shrink to "only small partial success" – quite understandably because every plastic particle in the ocean, every additional ton of carbon dioxide, and every species of animal wiped out seems to confirm the big picture of gloom and doom. In addition, within the community of sustainability activists, a David mentality prevails; Goliath are the others, optionally industry, overconsumption, the world market, the mainstream, politics or those up there.

Climate protection and sustainability goals are based on scientific facts, the elaboration of which is demanding and comprehensive. In contrast, it is comparatively easy to claim something is all-around wrong and to question successes with arbitrary assertions and without data. We know this asymmetry from climate protection. The asymmetry also occurs, for example, in debates on development economics, migration, or species conservation. It always depends on the context that sets the framework for understanding success and failure. This is a fundamental attitude, no matter how power is actually distributed in concrete cases.

It is important to understand this and to be able to communicate it well. This is challenging. Out of this relative positioning, successes can hardly be perceived, even if they are quite obvious to third party observers. Successful changes are all too quickly seen as normal, or they are deliberately sold as political normality. That is why around success nothing is self-evident. Thus, in order to register success as such, it takes mindfulness and an attentive spirit.

10.3 Tacit and hushed, but more important

Examples of failure are clear and obvious, for instance, when politics does not deliver against its own publicly declared goals. In Germany, for example, greenhouse gas emissions are not sufficiently reduced, the number of obese people is rising, landscape consumption is not being reduced, and the conversion of agriculture is lagging behind requirements.

But the more subtle sources of successes are the ones that bear greater importance. They are usually not talked about, which is not good as this non-action appears to add a strategic mismatch. As compared to the clear and open context of determining goals those subtle successes sometimes are having much more relevance to the overall context. They are the tacit and hushed agenda of sustainability policies. Opportunities may be tackled, chances may be recognised – far beyond the hustle and bustle and the tight race from transformation conference to transformation conference.

There is, too, a striking example of the double face of success, the realised and the missed one: Countries such as Poland, Australia and India in particular are holding on to coal as an energy source, because of the path dependency of their industries, but also the fact that they see no alternatives to coal. In the early 2000 years, the technology of carbon capture and storage (CCS) had just become marketable and offered a straw for getting away from deep coal and thus an alternative that would provide more climate protection. Germany, at that time, was among the technology leaders. Success in climate strategies seemed palpable. Plant engineering industry was enthusiastic about the solution that would enable it to capture the carbon dioxide in its plants, knowing that this solution would, of course, only be partial and that the main impact for combatting global warming would have to come from renewable energies. They left the second part of the solution, the disposal of the carbon dioxide, to the public sector. The one's were talking about high-tech and making money, and the others about cemetery technology and spending money.

As a rule, this division of labour does not work out. In addition, the German government made a remarkable miscalculation and presented a draft law for the approval of CCS plants that exactly reproduced this division of responsibility and depicted it as if on a platter. Burying carbon dioxide may be technically safe, but the throwaway ideology is out of touch with the general feeling in society. Utilisation of the sequestered carbon was not thought an option. That is why basically the whole approach failed in Germany back in the day. Twenty years passed idly by. Germany lost the technological leadership. Only now is there renewed commitment and investment in Germany because the carbon management of steel and cement industries and others is running out of options and there are physical limits to driving efficiency and reducing carbon emissions further, while the regulated market for emissions allowances demands ever higher prices. Carbon capture, utilisation and storage (CCUS), is entering into practice, for example in the smelting of green steel.

10.4 Getting the language right

Language is a success factor of utmost importance. When you get a story right and can tell that story in a way that the audience can follow, success sets in. Unfortunately, the reverse is also true, and this is often the case in the sustainability community. The use of language that is difficult to understand is widespread. This is a phenomenon because this community, in particular, is very engaged in educating people about all sorts of scientific findings, such as planetary boundaries, the effects of climate change, plastics in the ocean, or resource depletion. However, the 2030 Agenda is full of opaque wording that gives the impression that people are hiding behind obtuse language. The right language is the one low-hanging fruit for more tangible and participatory policies. Seeing this point is a gateway to success.

How people express understanding and interest, and how they are told things changes their mindset. The science of semiotics knows interpretive frameworks representing the mechanism of understanding within a cultural group. Social codes also include unwritten codes such as individual body codes and behavioural codes. One may wonder what goes wrong when language is not used to its full positive effect, which is often to be observed in our context. The key statement that we hear in connection with global warming appears in obfuscating language. It claims that climate change has caused an increase in natural disasters. No, that is an incorrect statement. Language is mixing subjects and described phenomenon. Climate change is not, humans are the subject. They are causing issues through their actions and inactions. Since the beginning of industrialisation, another key statement said, the Earth has warmed by one degree Celsius. No, the Earth has not done that, man has. Or a final one, from a myriad of language examples: a cross-sector transformation is required for the necessary change. Here again, this passive sentence construction obscures the subject of change, namely man.

Sloppy language is obscuring the real circumstances and constructs a wordy environment of its own that alienates addressees from reality. This may happen against the intentions of those speaking. The point here is that any message gains persuasive power from reality. The more a sentence corresponds to reality and to the real-life experience of people, the more likely it is that the message will reach out and convince people. Which is, of course, a means to success.

Something else comes into play. For obvious reason, superlatives and alarm statements are increasingly being used ("it is five before noon", for the last three decades an often-used framing). But the social budget for the use of superlatives and alarm statements is limited and can easily to be overdrawn. Closed circles of protagonists and media increasingly use an almost homogeneous language when talking about global warming and unsustainable trends. This language is meant to stir up the public, but one may doubt that this really works out. It may even prompt the opposite. Passive language doesn't necessarily speak to people. And if it does, it loops people into passivity.

People interpret the same information they receive in different ways, depending on the values they hold and the life situation they find themselves in, whether social, political, cultural, or religious. This is a factor for success, if properly understood and used. In addition, fear and alarmism can easily come across as manipulative, where as a positive story about success can encourage or empower the audience.

Fatigue, as, for instance, in dialogue fatigue, climate fatigue, is an attitude that we observed ever so often. One-sided stories about crises or one-sided stories about imaginary brilliant-to-sell success do not help if they use only passive voice that leaves in the dark who is actually doing what, it can leave people feeling overwhelmed, powerless, and without an alternative. Passive voice generates no response outside the bubble. The passive sentence is programming for failure, the active sentence is programming for success. The passive voice is the language of failure, the active voice is the social code of success.

In the context of this book coding for success is an institutional and governance feature rather than something individual. Powerful and clear language codes affiliation, status and, acknowledgement and expertise. They engage people and encourage sharing of experiences.

10.5 Think big and start small

When implementing a big idea, the smallest possible step makes the most sense. Big implementations are complex, and windows of opportunity open and close at a rapid pace. Therefore, each step, however small it may seem, should be placed in a political cycle that manages organisational learning. This prepares organisations to gain experience and view their own approach as iterative. Small test balloons or idea bombs will find that many people are already well-prepared and ready to act, so "only" the right framework and action design are needed to trigger full action. Wherever you start a project, whether "at the beginning" or at a mature level, you should plan for your options to finish the project, or better yet, to fit it into existing structures.

Starting small and progressing step-by-step is a necessary, but it is certainly never meant to be sufficient. Small steps need to be embedded into a larger picture that enables its protagonist to provide reliable answers to questions like: For exactly what purpose are we doing this? Who should be our target group? How big are the competencies that the purpose requires? Don't we overload the purpose with expectations? How can we avoid that the purpose unintentionally excludes people or organisations through indirect consequences that are not openly communicated (causing real problems)? How can ordinary people audiences interact with the purpose? Does the purpose use appropriate and friendly language? What kind of narrative does the purpose represent?

Starting small is a big deal. But small needs background.

Any achievement quickly sinks into oblivion if one does not know the root reasons that led to success. There is no real understanding of self-efficacy without some idea of the historical roots or on which shoulders one stands while reaching the next level: who pioneered the movement, who made the basic scientific breakthroughs, who then helped scientists turn basic science into applied and communicated solutions, who irritated policymakers and by what means, who thus initiated a rethinking that started the shift in consciousness, who and what has enabled science to research the Earth system in the first case?

The Fridays-for-Future generation is learning in school about climate change, plastic in the ocean, food waste. The actual discussion tends to take the elementary facts about carbon emissions and key concepts for granted, without knowing the historical background, much less the causes of change we've seen in the past. For the media and activists, the climate battle begins with the success of Paris, i.e. the international climate agreement reached in Paris in 2015. The contradiction between this ambitious decision and the subsequent failure of "business as usual" is considered to have triggered the Greta-like outrage and the recent career of the issue. This short-term horizon is the backdrop against which all actions are judged as success or failure. This may be good for public campaigns and media storyboards. It is totally inappropriate for dealing with the problem and transforming the world.

Sustainability action identifies itself via negation of big trends, not via success along small steps. This is one of the defining elements of the sustainability movement in Germany, and may be all over the world. Interestingly, this runs contrary to daily life experiences e.g. in sports or at school where success and repetition of success coins the key learning experience. The relation of What (goal, targets) to How (steps, implementation) is an intriguing element of successful strategies. We can differentiate three cases:

- The what derives from the how.
- The what is soft coupled with the how
- The what and the how are of equal value.

Starting small means working in a place where you're in control, and showing that you still mean business and that you're willing to push the boundaries of normal. For instance, you may show that you are the one who acts as a role model by not wasting food, moving away from fast fashion, being mindful in the way you consume and use resources. Or taking the initiative to start a neighbourhood recycling club, green catering or discussion meeting. In a company, getting started with sustainability can mean taking the first steps toward reporting or compiling a list of what is already being done in the various departments and what the company's vision is or could be. Those small steps will create momentum and get people going. As people may find first actions encouraging or even adventurous those actions may free people's imagination. The success lies with the next step. The what (to do first) is pretty much identical to the how (to generate interest).

For businesses and large organisations, the small level has a different dimension. It is tempting for companies to strive for so-called Big Hairy Audacious Goals. According to Jim Collins and Jerry Porras (2004), who introduced this concept, this is about a compelling and easily communicated goal that an entire organisation strives to achieve. It is complex and intriguing enough to inspire employees and spur them to action, even though, or rather because, it offers an irritating big picture vision that, in the beginning, no one knows exactly how to implement or exactly when the goal will be achieved. Implementation, the message goes, means inventiveness and ingenuity, not just following prescribed rules. This particular out-of-the-box way links the what (to be achieved) with the how (to proceed).

In public administrations, at least German experience shows, neither approach works without obstacles. Both would foster those administrative energies that end up avoiding action that conforms to goals or steps without openly saying so. Or they would overstretch commitment through bureaucratic management and control of silos. To be successful, public administration needs orientation through democratically legitimised goals. Executive goals, no matter how high they are on the commitment scale, are not good enough. Operational goals or intra-institutional purposes alone will not set the organisation on the right path although this is sometimes hoped for (European Commission 2017). What really helps is some kind of guided goal process, where the overall goals are set by democratically legitimised institutions (mostly parliaments) and where those institutions or auditing bodies review the performance of the administration, and quite necessarily, advance the governance (Meuleman 2021). The point here is that the what and the how are of equal value.

Finally, the as-if-success needs to be mentioned. It is widespread. Contrary to popular belief, the high number of participants at international or UN summit conferences is not a factor in their success. It is a major misconception: the public, and of course the participants, read a high number, such as the 25,000 at the climate conference in Glasgow, as a sign of significance and power. In reality, however, only a few hundred people are actually involved in the negotiations; a much smaller group of them finalises the agreements. The large crowd of bystanders from academia, non-governmental organisations, advocacy groups and the media mostly wait it out, present their studies and network in so-called side events or act as commentators. They call this "applying pressure on negotiations". However, success was never measured. Actually, it manifests the as-if-feature: as if crowds of bystander matter, as if travelling means commitment, as if such summit meetings can ever be a redemption.

Again, the roots to success lies in one's own action in the domestic sphere, however small, because those with political responsibility are directly called upon here. This will then also be transferred internationally.

10.6 Transformation, mindshift and success

Transformations, even great ones, are not unique in the history of mankind. The current transformation, the Great Transformation, will be no exception.

Historically, transformations represent a mixture of continuity, decline and new formation of social reality (Watson 2010). Medieval Europe saw several scientific transformations of thought, social group formation, technology, and governance. The universities that emerged at that time cracked the magic monopoly of the church and feudal rule.

Gutenberg's invention of printing technology started a transformation that can be called dramatic. The decisive factor, however, was that his invention met with a receptive society. It became the social norm to read silently (i.e. to dispense with the reader reading aloud). This made it possible to read faster and more, and more people did so because more reading material was available. The demand for eyeglasses increased. The craft of eyeglasses flourished and was extremely innovative and inventive. Precision engineering led to telescopes and the microscope.

A look at history sees a society that doesn't look like centuries past at all, but can be described in modern terms: its innovation is open to technology, promotes competition and focuses on knowledge and education. It brought about special financing concepts, developed the first forms of state governance and allows for complex interactions (read: network functions) between these features. In the Renaissance, craftsmanship and technology (iron, copper, wire, gold, silver, hydropower as a source of energy) interacted with demographic development, the spread of capitalism and the emergence of statehood. Knowledge, specialisation, civic culture and science created the ink state. Culture, politics and social relations are at once the source of innovation and its condition. Trials, experiments, coincidences and aberrations, successes and failures abounded. They form a colourful and diverse pattern. But recognizable are also the basic features of transformation as expressed today in contemporary categories: Purpose, scientific (technical) inputs, goals (then related to authoritarian power), networks, craft and practical expertise, political (then mostly financial) cycles. A so-called error culture (error-friendliness) is known only in modern times. But at that time experiments enforced the competition of good ideas against regressive ideologies and power (inquisition). All great figures are associated with the fact that they left the path of rigid convention at a certain point and dared something new and unfamiliar. They went off-script.

What does this mean for sustainability policy today? What factors are driving success?Education, openness to new technologies, and competition of ideas are equally key concepts today. They are just often undervalued or assigned to political conservatism. They are often met (in Germany) with fundamental skepticism, as in recycling, the bioeconomy and genome editing, the production of hydrogen and innovations in the food system. The cultural dimension with which these techniques can be scaled is little addressed and hardly used. Digitisation is held back by weak infrastructure and the lack of a sustainable digital policy.

The pentagon of transformational practices (see Figure 6.2) is the framework for success. At each position, the factors mentioned are important for success: step-by-step approach, active language, reality fit, networks, looking at what is hidden.

Sustainability strategies have done too little to address these issues. They could, however. To do so, their sponsors would have to adapt the contents of the success pentagon instead of sticking to the myth that the main work is already done with the establishment of goals and that everything else is only a question of the "means of implementation", i.e. essentially political will and money.

Of course, it is tempting to proclaim a decarbonisation that relies entirely on the service economy and simply abolishes industrial production in unison or pushes it abroad. However, this is a dead end. The transformation requires more rather than fewer industrial jobs because the solution often lies in new processes and high-tech products. The circular economy, an energy upgrade of our cities, the transformation of roadways, the hydrogen economy and the sustainable food economy are primarily associated with more work. This work places higher demands on qualifications and knowledge management; and at all levels, from the supervisory board to the most menial jobs. This is true even with less growth. The solution is not to move away from industry, but to carefully figure out what has an impact on the future, what makes a difference, what the next step is, how to take stock, and how to convince people.

Talking about such big aggregates as "the" transformation, "the" capitalism, "the" growth, or "the" economy is often too superficial. In any case, one must always be aware that these aggregates have very different manifestations in different countries. The real range of these manifestations is much wider than the term suggests. Its uses reflect a certainty that is deceptive on closer inspection.

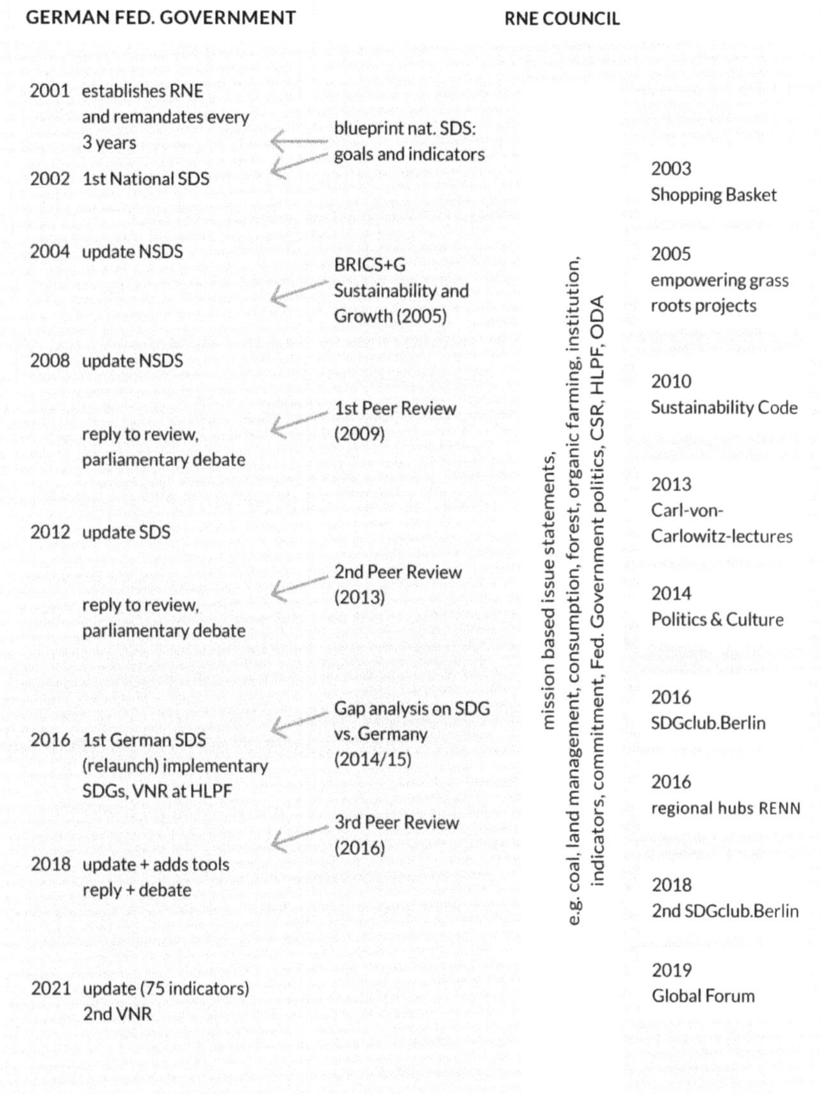

Figure 10.1 Operative interaction between federal government and RNE

More interesting is the question of when and how a company and an entire economy lose the "license-to-operate", i.e. when they delegitimise themselves or fail to offer a meaningful and credible option for the generation of grandchildren.

In Germany, the family business leader Christiane Underberg introduced the latter as *Enkelfähigkeit* (suitable for grandchildren).

The same applies to professional consulting activities and dealing with political cycles. Here, the important part of the license is implicit and rarely mentioned explicitly. Mostly, of course, there is a formal mandate and a formal code of conduct. However, political practice is built on trust, reliability of deadlines and response times (regardless of the content of the response), convening power, and a willingness to engage. The ways in which the Council for Sustainable Development so far intervened in policy, produced role models and new networking approaches, or made timely policy recommendations provide examples of how the informal license to deliberate can be interpreted beyond the usual blame and shame games (see Figure 10.1).

The governance rules for sustainability are still in the works, and they will probably never be complete or finite. Despite the new and sometimes surprising challenges posed by the rampant problems, governance is still restrained and has become deeply ingrained in the attitudes and habits of those in charge. In the German case, the emergence of new approaches and the adherence to good old textbook solutions balance each other out. Here are the most striking ones:

- Purpose

The concept of "purpose" comes from the world of private enterprise and marks a significant shift in the corporate world. Making money, capitalists said for decades and still say, is the primary, if not the sole, purpose of doing business. Most of them still say that today, while others tell a different story, at the core of which one often find the term purpose. For them, the license to do business is more than just money. At the heart of their purpose is their responsibility to society and the environment. They view their economic activities as a contribution to the environment and society at large.

This concept applies to any organisation and is not limited to the corporate world. In terms of sustainability strategies and climate change policies purpose in this broader sense is not simply given by quantified goals, even if they are expressing high ambition. The concept of purpose points out that those quantified messages may even counteract the purpose if they are treated more like a political trophy than a true guiding principle (which, unfortunately, is often the case).

In this context, a purpose is more than just a set of quantified goals and timelines. It is a core belief of success that can be expressed through a vision or a mission statement. Finding the purpose and putting it into words certainly means finding the purpose again. There is little point in "inventing" a purpose from scratch. Even working out a purpose in workshops or through design thinking will not necessarily bring the desired success. This is mainly because the purpose is already there, but buried deep in everyday routines and tacit knowledge. It needs to be brought out again. Thus, in most cases it is about retrieving instead of inventing the purpose.

Promotors of a purpose concept need to translate it into a real-world to-do list. But it also needs a virtual space to experiment with its relevance and to gather the resonance the purpose generates in the eyes of external observers, customers or everyman voters.

- Networking networks

Italy is a good example of networks. The Italian Alliance for Sustainable Development (ASviS) has been in existence since 2016, founded by foundations and universities, and increased its membership to 270 organisations in June 2021. ASviS organises numerous actions for school education and public awareness with a strong cultural and political dimension. So this network has is a clear message. Civil society organisations and businesses want more and demand better political and administrative frameworks for sustainability. As a meta-level network (a network of networks) ASviS provides a legitimate and practical basis for organisations. At its core is a concern to improve government action and accountability.

In Germany, there is increasing talk of the *Gemeinschaftswerk*, a collective network of networks designed to drive the implementation of the sustainability strategy. As early as 2011, the idea of a collective community effort was floated in the debate on Germany's energy transition, but never implemented (see Chapter 5), much to the detriment of the energy transition. Chancellor Angela Merkel and the minister presidents of the 16 German states agreed a new Gemeinschaftswerk concept in 2019, along with a number of ideas on how public procurement could take the lead. Implementation is currently prepared by the German Council for sustainable development.

In other German contexts, such as multistakeholder consultation on the failures of modern agriculture in terms of biodiversity, animal welfare, or sustainable finance, the practice of linking silo networks has proven to be the way to go.

- Making it a personal challenge

Institutions tend to organise accountability for sustainability anonymously. It is then often said that sustainability is cross-cutting and that everyone contributes. That such contributions need to come from each individual work unit is of course undisputed, but as a governance principle it is inappropriate. This is why so many well-intentioned approaches to "integration" and coherence are so ineffective. An evolutionary progression is to address ownership so that sustainability has a face, a phone number, and a social media account. Among the tasks thus assigned is coordination. In an even better case, coordinators are grouped together and centrally managed. They would have a shared budget to draw on for ideas and innovations. Proportionate equal distribution of the budget among all participants is out of the question. Rather, winning a grant from this budget would be a matter of personal commitment in the symbiosis of competition and shared responsibility.

The task at hand is coaching. Sound coaching is something that helps leaders and activists at all levels, even if they think they know everything and have access to extensive resources. This coaching is not just about hard skills and soft behavioural skills, a healthy work-life balance and mindfulness (although this is all important). Coaching for sustainable development is much more. It's about a mindset that enables people to stay on top of their game and fit for the purpose at hand. When working in the field of sustainability, it is imperative to continually explore one's options and reflect on what seems right and wrong and what may be overlooked in between. It is about the ability to be surprised by solutions, by other views and own weaknesses. And it is about the awareness of the crucial reality of runaway problems without losing the courage and faith in the ability of people to change the course of life for the better. Success and failure can be close together, and one must be able to continually learn through mutual exchange with peers and collaborators or others. Creating and using language that fits the goal is another important point in coaching. The language should be understood not only in narrow professional circles, but also by the public. It should make it possible to "sell" visions and at the same time draw attention to new developments. This is what one may introduce as sustainability skills (as an addition to hard and soft skills).

Megan Darby (2019), editor of *Climate Home News*, tells another one of those almost unknown success stories that are, yet, of paramount importance. It is the story of the Lionesses of climate policy. In 2013, this group of about 30 lawyers, diplomats, financiers and activists met informally at a country estate on the Scottish border because the official climate diplomacy seemed to be stalled and not getting to the point. Those women invented the storyboard for the Paris Meeting and the way toward it. In particular, they coined the idea climate neutrality and decarbonisation and developed (and implemented) a plan to bring those ideas into real politics. They made a dysfunctional international climate discourse a personal challenge, and were successful.

- Dancing the governance

The institutional architecture is the be-all and end-all for long-term and consistent success. This is where essential expertise lies, provided institutions open up to the topic of sustainability. All countries are looking for ways to achieve successful transformation. No one wants to set up completely new offices and control authorities, but the old ones have not really worked so far either. Only a few countries are refraining from political and organisational innovations altogether. Some impulse toward governance is taking place everywhere. The central question is how to fill the role of gatekeeping. This refers to an institution or a person who exerts central influence on strategic decisions while credibly representing the concerns of practitioners. Of course, governments are committed to involving as many stakeholders as possible and to "leaving no one behind", but institutionalised forms of socially relevant consultation have only been established in a few countries. Certainly not commissions with representatives of society or Sustainability Councils with budgets, their own agendas and equally strong reach into both government and society. For that would require governments to trust civil society expertise to a degree that is alien to their political culture.

Silos (communities of expertise) should not be "broken up" anywhere just because they are silos (see references in the chapter on governance). When they function well on their own, they can be helpful and are often indispensable. Operative functionality is an important part of the executive governance of sustainability strategies. Legislative, parliamentary governance needs to be developed around compliance of draft legislation and budget authority.

- Science based operations

The question is old, and never more up-to-date than in contemporary times: Should one base policies on evidence or should the respective evidence govern the kind of policy response that is needed? Science-based policies or evidence-based policy? The practical answer to this question is no longer at the discretion of policymakers alone. As humans are threatening the human life natural support system the appropriate response becomes more global and more complex than ever before. Decisions about the future have become even more important to the public. Governance for sustainability needs to come up with a concept and framework for how to make the best use of scientific knowledge and how to deal with uncertainties.

- Leading by example, walk-your-talk

Managers and employees are trained together on sustainability. Technical content and procedures are exchanged and further developed to the best of our knowledge. Requirements for third parties or the market (for example, on quotas for sustainably produced food in canteens and cafeterias, fleets of electric cars, priorities for rail versus air travel) are mirrored in the company's own administrative area of responsibility. Green Procuring's high demand for information and training is met with digital efficiency. A sustainable event management (healthy food, no food waste, carbon footprint, paperless, etc.) proves to be an excellent entry point for developing a walk-the-talk routine.

- Learning, organisation-wise and as society

Organisations that empower themselves to engage in continuous learning are better positioned than others to support and drive sustainability practices.

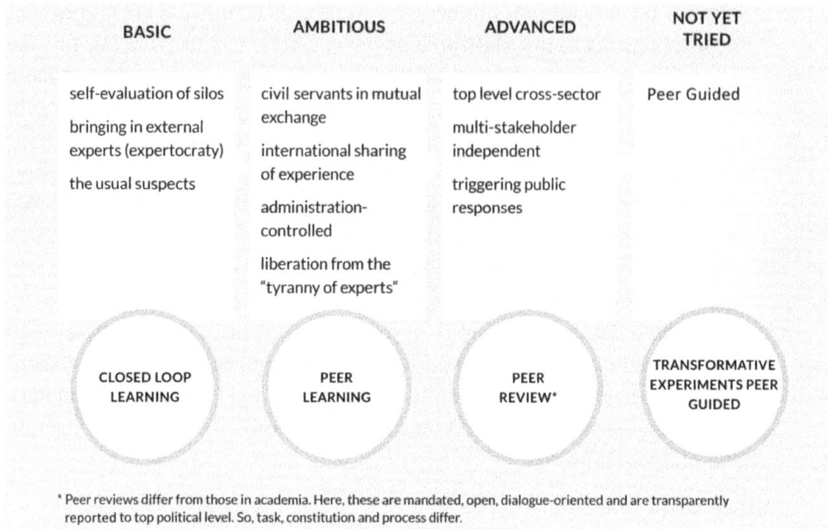

Figure 10.2 Organisational learning in the SDG context

> The same is true for entire societies, although societal memory is a very complex issue. Societal learning and forgetting take into account culture, routines, and other important social mechanisms of continuity and renewal in human communities and sociotopes, such as post-disaster trauma, both individual and collective, or post-event stories of successful transformations. Learning from experience organises the influence that the past has on the present behavior of a given subject, keeping in mind that the present simultaneously shapes the understanding of the past.

In the context of sustainable development, learning mechanisms are primarily tied to organisations. This is because the implementation of the SDGs is primarily subject to an organised form of national decision-making and governance. The processes of gaining, sharing, and transmitting insights, knowledge, and big data improve implementation over time. Engaging multiple stakeholders in an advisory and platform capacity is key to strengthening these processes, as it brings in different perspectives that shed light on areas that might otherwise not be in focus (See Figure 10.2).

There are a number of fundamental characteristics of organisational learning. When managed properly, they are fit for purpose and add value to policy and implementation. Here, organisations can use internal feedback loops and internal auditing standards. They often stick to the fairly common routine of bringing in selected experts or consulting firms. This, too, can be effective, although some call it tyranny of the experts. A more ambitious feature seeks to free organisations from dependence on expertocracy. So-called peer learning, which has only recently been introduced, is used for the mutual exchange of governance knowledge among small groups of administrators from different countries. This learning is internal and sectoral (within traditional administrative bodies), but now also cross-administrative and across different policy frameworks. Even more advanced, however, are peer reviews that cut across countries, sectors, and societies. They are open and transparent. This feature engages prominent multistakeholders from different countries and backgrounds. Because of their personal status and capabilities, they are expected to represent independent views and address uncomfortable truths. Whereas peer learning is a decentralised feature, peer reviews are more likely positioned at centre stage of policy-making.

As a next level of learning, one might consider real-time transformative experiments led by peers. This feature is not yet implemented anywhere.

References and recommended reading

Collins, Jim and Jerry Porras (2004) *Built to Last: Successful Habits of Visionary Companies*. New York: Harper Collins.

Darby, Megan (2019) *Net Zero: The Story of the Target That Will Shape Our Future . . . and the 'Lionesses' Who Changed the Way We Measure Climate Ambition*, www.climatechangenews.com/2019/09/16/net-zero-story-target-will-shape-future/

European Commission (2017) *Quality of Public Administration. A Toolbox for Practitioners*. Brussels: Publication of EC Employment, Social Affairs & Inclusion.

Meuleman, Louis (2021) *SDG 16, Public Administration Reform and Success Factors, 2020–21 Executive Training for Policymakers on the 2030 Agenda and the Sustainable Development Goals (SDGs)*, https://unosd.un.org/sites/unosd.un.org/files/session_4_mr._louis_meuleman.pdf

Watson, Peter (2010) *The German Genius: Europe's Third Renaissance, the Second Scientific Revolution, and the Twentieth Century*. New York: Harper Perennial.

11 Seeds

The evolutionary germination principle, repurposed

11.1 The seed parable: a borrowing from nature

In ecosystems, seeds are a key element for evolution. They organise reproduction of species and success of life. In the human context seeds are an element for success, far beyond mere technicalities, money, or governance, but not without them. Seeds are a symbol for hope, trust, and for the willingness and power to survive. Politically, the seed is the parable of how a platform, an external condition, a general trend, or other given circumstance can cause ideas to sprout or raise the right question. Seeds symbolise the willingness to explore the past and the future and make the most of it.

In the context of climate change and progress toward sustainability goals we see quite a lot of so-called theories of change, with partly large communities of followers. The business sector as well as philanthropies, NGOs and administrations sometimes use theory-of-change approaches for the planning of the social change they want to see happen. In that sense, although seldom explicitly expressed sustainability strategies, too, are examples for implementing theories of change as they are defining long-term goals and design policies to achieve them. The theoretical idea behind this that a goal allows mapping backward and that this exercise leads to identify necessary preconditions, an insight that would then be used to redesign goals or to reshape the implementation. Backtracking usually needs to suggest that change processes rely on causal linkages or rational frameworks. Thus, theories of change certainly have their justification. Here they are only recalled to make clear that the seed parable goes a different way that maybe even end up in a change of theories.

In our context, a seed is a potential for action and transformation that is in an embryonic stage and may even be surrounded by a protective shell that makes it difficult to even recognise the seed. It may well lie dormant for some time, thus surviving times and external conditions that are unfavourable. In our case, seeds can be dispersed by environmental conditions, much like natural seeds are often dispersed by wind, water, and animals. Not all seeds will germinate, as germination depends on proper growing, acceptance, and cultural conditions. Of course, seed viability is key. It determines the proportion of ideas and initiatives that eventually take hold and make a difference. This is exactly where social

DOI: 10.4324/9781003242024-11

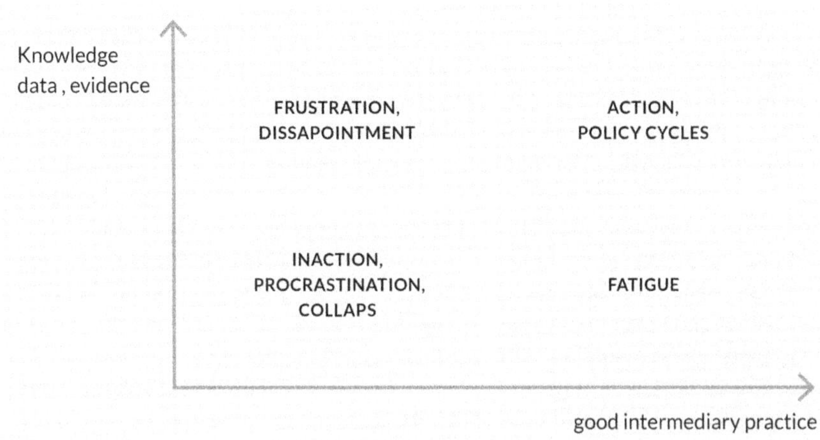

Figure 11.1 Four types of impact

intervention comes in. The parable of the seed offers a practitioner's approach to change. Instead of designing and predetermining pathways, it chooses from what is available. Instead of examining causal linkages, the seed idea simply looks for entry points, no matter whether they are hard-wired (causal) or soft-wired (cluster), small or big.

The famous French poet Victor Hugo, elected member to the Académie française in 1841, is noted as having said that "nothing is more powerful than an idea whose time has come". This was a big word then, and it is still today. Hugo did not say that change is happening right away. The connection between idea and time are people. And they react very differently. Change is not the same for all people. Some are afraid of change. They withdraw and try their hand at inaction. Rather, they may prefer the known discomfort to an unknown risk. Incidentally, the same can be said of those who criticise the system but are satisfied with their comfortable opposition role or even make it their distinguishing feature.

Another group may feel exhausted by the flood of scientific data, theories and approaches because, as it appears to this group, they revolve too much around themselves and have too little connection to real life and actual effects (see Figure 11.1). The frustrated, on the other hand, think they have convincing arguments. After all, they see themselves in an alliance with science and the prevailing scientific opinion. They are disappointed by politics, which, in their view, does not do enough.

The fourth group includes people who use data and information in an enlightened way and at the same time also choose skilfully between the options for action. Effects emanate from all four groups. Of the latter, they are most associated with confidence, satisfaction and creativity. From this group the attention and the understanding can be required to see and use the seeds. That is another reason where multistakeholder bodies play an eminent role in discussing off-script procedures and encouraging new thinking.

Only the fourth group is in a position to potentially ask the questions that are the right ones in the Anthropocene and that are so new and unfamiliar. Normally we know how to deal with the social issues e.g. in terms of the social security system, fair wages, and respectful taxation in response to the Social Question raised by the Industrial Revolution in the 19th century. This is a people-against-people struggle. People are struggling with people to find solutions for people. Here, interchangeable interests of subjects are measured and negotiated in terms of money and legal rights. In many countries, this is still the serious first and most important issue societies deal with today. As the dominant and primary important mechanism, however, the Anthropocene puts it in the museum. Now it is a matter of humans versus nature, of subjects versus objective circumstances. The negotiators have to deal with objective realities (resources, risks and damages, scarcities, transformation) that do not speak for themselves and are to be legitimately brought in by proxy. This requires a completely different approach to problem solving. Previous chapters already made the points on shifting governance features, on gatekeeping institutions and the politics of strategies, and on the culture of success and failure. Here now follows, in concluding, are some of the more general approaches that might be of even higher interest for advocates of transformation in other countries. They are best expressed by the concept of seeds, as explained in the following.

11.2 Seeds of hope

Sustainability is the defining issue of our time, which, as far as we know, is rightly called the Anthropocene. There are two landmark sentences that could be not more different: Martin Luther King's "I had a dream" and the "how dare you" of Greta Thunberg. Both sentences are landmarks of their time, and both express different cultural framings of hope: the former inclusive and encouraging, and the other, desperate.

In recent decades, science has repeatedly alerted us to the dangerous, human-induced effects of climate change with increasingly ominous superlatives. Climate scientists continue to work to deepen their understanding of how the Earth's climate system works. Today, however, open eyes and an open mind are already finding confirmation of the science-based warnings. Effects of unsustainable pathways can be seen in almost every forest, every landscape, in our cities, and literally in everyone's backyard. This is the backdrop of the changed and now increasing public use of the term sustainability. All the more reason to ask for ways out and alternatives, and how all this relates to mindscape of people, for which hope has a special role. Hope is the one resource that can fuel up solution finding, and the least researched.

The 2030 Agenda and its 17 sustainable development goals refrain from touching on hope. Those who believe in new technologies believe that big technological fixes will solve the problem. Those who believe in innovation believe that innovation will save the world. Those who believe in campaigns and protests believe that demanding great goals and the voices of young people will change the course of life. Those who believe in green growth believe that a greener world will overcome the dangerous effects of climate change. Those who believe in bold goals and binding legislation believe that those will determine implementation. The same pattern holds for belief in competition, emotion, imagination, ancient survival instincts, modern scientific knowledge, or religious faith.

Each of these beliefs is valuable and they do express hope, even if it is only appearing in small portions. They each have a rational background in some way. Their promises are not totally false; however, they are incomplete because of their intrinsic limitation. Their input assumption determines the outcome of the reasoning. Mostly, those argumentations follow their own, pre-constructed cluster, which is impersonal in nature and completely parallel, meaning they are carefully avoiding to take reference to each other. They like their own way and are driven by that way (growth, innovation, protest, etc.) without even mentioning interdependencies in the way of synergies or conflicts. They offer tunnel visions that are reflecting the mentality of a given community, and so they find themselves end up preaching to the converted. Experience shows that this makes it hard to reach out to other communities of the otherwise converted.

Since the dawn of philosophy, thinkers have devoted considerable attention to the social workings of imagination and hope. Modern psychology and political science provide theories of hope (Snyder 2002; Rand and Cheavens 2009; Porritt 2020; Homer-Dixon 2020a, 2020b), in which hope is defined as an individual's perceived ability to achieve his or her goals. It includes the motivation to pursue alternative paths and the ability to follow through on those paths. Specifically, it means finding ever new ways to incite the fire rather than just hailing the ashes. It does not matter whether this ability is learned or otherwise acquired, whether it is a result of self-esteem, or whether it is simply part of the mental infrastructure of a group.

Hope is a positive anticipatory emotion in a certain context. It plays an important role in psychotherapeutic treatment, e.g. in cancer therapies and in the rehabilitation of injured athletes and people in general. Hope is also a societal category. In the case of society's capacity to create sustainable development, hope lies in the conversation between mind and power. It is intrinsic and audacious. This hope does not arise, as it might in other contexts, from ambitious goals alone. Hope does not result of the simple equation of willpower and skills. It rather follows out of understanding of root causes, trust into solutions, and the feeling of not walking alone. A good sustainability strategy would produce and encourage hope by showing what works and what does not, or what can be made to work.

After Germany's liberation from fascism and the end of World War II the public opinion once called a speech given by an American The Hope Speech for Germany". It is still historically important. Officially, US Secretary of State

James F. Byrnes gave it the title Restatement of Policy on Germany on September 6, 1946. Many Germans were hopeful that Byrnes outlined the procedure for establishing a democratic state in Germany as a whole and that he believed Germany should be given "under proper safeguards . . . primary responsibility for the running of their own affairs" (quoted according to Time 1946). However, the Cold War put this hope to rest. In West Germany the social silence about the guilt of fascism took over and the Wirtschaftswunder (economic miracle) replaced any leftover of hope with heavy consumption. The idea of a democratic awakening was postponed until its tentative return with an election campaign slogan. In 1961, the Social Democrat Willy Brandt described an important demand of his programme for the election to parliament (he won the elections a couple of years later):

> Shocking research results show that in connection with the pollution of air and water, an increase in leukemia, cancer, rickets and blood picture changes can be observed even in children. It is dismaying that this community task, which involves the health of millions of people, has so far been almost completely neglected. The sky over the Ruhr region must become blue again!

What was then the blue sky over the Ruhr (one of the most industrially polluted areas) is now sustainability with a special focus on climate neutrality, although they obviously do not have such a strong image power as the blue sky over the Ruhr area.

US President Barack Obama (2006) coined the phrase "the audacity of hope". We in Germany like such pathos when it comes from abroad. The German language of political normality is rarely capable of it. That is maybe, because the politics of hope formulates a blatant and unabashed imposition on everyone. This type of hope is about self-empowerment and a freedom that strives for meaningful action rather than merely being purpose-driven. Beyond the normal solidarity of the like-minded hope catalyses when the differences and the different are brought together. Dialogue, in that sense, is a hope producing machine. Wild and hopeful thinking can help a society see itself more clearly and take sharper action. So, hopeful doing for oneself, for a group and for the society at large may liberate people and may make them strong.

This is the positive side of hope, but unfortunately it also has less positive and even negative sides when hope is being disappointed. Ignorance, frustration, rejection and sullenness can be the result. All too often insensitive exaggerations and hypes degenerate hope into self-righteousness and blind egoism. Crushed hopes echo in every generation everywhere. Barack Obama (2004), obviously aware of the problem of crushed and fizzled hope, calls hope the belief in simple dreams to demonstrate a more concrete approach. In his memorable speech at the Democratic Party's nominating convention on July 27, 2004, he enumerated what this faith in simple dreams, this insistence on small miracles include: that we can tuck in our children at night and know that they are fed and clothed and safe from harm; that we can say what we think, write what we think, without hearing a sudden knock on the door; that we can have an idea and start our own business without paying a bribe; that we can participate in the political process without fear of retribution, and that our votes will be counted at least, most of the time.

Anywhere people have those simple dreams. They cannot be simple enough to form the building blocks of sustainability as a common concern. No matter whether expressed publicly or hinted at in private, there is a need for humility in the face of the Anthropocenic magnitude of a task that is much greater than the day-to-day political struggle. It seems like too small a coin compared to the big currency with which the "political Berlin" and the national governments across the world post roadmaps, frameworks, visions, programmes and plans. The politics of hope can be much more substantive than those.

The United Nations Sustainable Development Goals tell a story about aspiration of a better future and about the universality of that notion. In principle, this is a good story that runs counter to the atmosphere of fear and conflict in the world today. And this story is not blinded by idealism or naiveté. It underscores the fundamental value of the immediate conversation of people working together in networks toward a common cause. The most interesting thing is that they might use all differences of interests and their being different as an asset. This ethics is a stronger political language than legal paragraphs or the regular political rhetoric. Making one's own responsibility effective is the strongest lever for the broad implementation of sustainable development. Mutual encouragement is important.

Sustainability policy is about more than just setting ambitious goals – not because that would be wrong, but because it often remains too abstract and remote. It doesn't come across as elementary as it is meant to be. It lacks a good connection to people's everyday lives and to the very practical reality of life. The 18th SDG is missing. It should have been on hope as a catalyst of change during a global period of growing polarisation, disintegration, and frustration.

In Europe, the political faith of the idea of sustainability also tells a story of hope. The two or three decades after World War II triggered a concept of seemingly unlimited progress and of unbroken administrative functionality. The 1970s saw the rise of institutionalised environmentalism. The first-ever UN Environmental Conference took place in Stockholm, a reference to early environmental pioneering role of Scandinavia. In 1972, Donella Meadows, Jorgen Randers, and Dennis Meadows (Meadows et al 2004) created a computer model that they used to analyse trends in global resource depletion. Their findings shocked the world and sparked a stirring discussion about environmental overshoot. The authors did not make dire predictions. Rather, they calculated the model in the hope that public discussion of the results would lead to them not occurring. The limits to growth argument gained its prominence because it formulated the hope for insight, enlightenment, and correction. Later, Dennis Meadows seems to have lost that hope, and in 2012 (personal communication) he said that it was too late for any efforts at sustainable development strategies because the world was beyond saving.

In the last half of the last century, the idea of political change boomed in Germany. It was hoped that the political antagonism between East and West could be overcome and at least civilised. However, instead of distributing a peace dividend at the end of the Cold War and after the collapse of the socialist states, globalisation ushered in an era of egoism, nihilism and mistrust. Neoliberalism permeated societies and political culture at large reducing hope to a category of egoistic self-interest. With the international banking crisis of 2007/8, growing geopolitical differences, the pandemic and recent climate-related disasters, this may now be coming to an end. The open question is which policies will be able to adequately shape the defining issue of the next decades.

Hope means business. That is why international consultancies regularly publish the results from surveys of senior business leaders on what they perceive to be the trends and developments. Sometimes several hundred to several thousand CEOs are surveyed. In addition, there are a number of central conferences ("Davos") that are considered places-to-be. Sustainability often plays a central role here, accompanied by the prospect of new business and business ideas. The number of companies engaged in circularity, climate, biodiversity, diversity, equity, and other sustainable development agenda issues is steadily increasing. According to the annual global survey by GlobeScan and the SustainAbility Institute (2021), many leaders of multinational corporations speak of a new momentum in sustainable development. The surveys 700 experts from 70 countries believe that sustainability is becoming embedded in more and more aspects of business and daily life.

There is little doubt about the correctness of this description of corporate insights. The most striking element of those snapshots is not the particular message, but who has been asked and who not. There is no equivalent survey among leaders of initiatives and grass root organisations devoted to sustainable development. But they are the ones who turn hope into policy, if that is the case at all. They organise those projects and campaigns that reach a broad public, and their word is considered credible by the media, whether always rightly or sometimes wrongly. Here are the practitioners who understand how to irritate the mainstream without losing the link to the mainstream. They are the ones where one may find the strategic intelligence to go off-script, to tell the stories of heroes that make others heroes and how to let your organisation (and yourself) be transformed by the hope for a better future, long before the future happens.

Future trends cannot really be determined without their insights and narratives. A sustainability policy that relies on, conveys and strengthens hope urgently needs a GSHS, that is, a global survey on the state of hope for the success of the 2030 agenda. Most probably, such a survey would discover the seeds of change.

In biology and ecosystem research the concept of dormancy of seeds is well known and often described. The same concept applies to seed ideas in politics, business, culture and innovation.

11.3 Seeds of arts and cultural practise

Art offers a special form to recognise reality and to express hope, and has always done so. In the Anthropocene, this becomes even more important. Scientific research and Big Data models provide policymakers with a reasonably solid basis for decision-making. The importance of science is undisputed. However, the role of art and culture is underrated, but no less central in challenging free thinking, and promoting hope and solutions. Truth is always a matter of ratio and of scientifically determined data as well as (!) of emotional images and interventions appealing to all senses. However, for the latter appropriate structures are still lacking. This is due to both sides, the arts as well as sustainability politics: artists insist on the so-called autonomy of art and defend themselves against a supposed instrumentalisation. Governmental as well as non-governmental actors have a clear instrumental relationship to the arts and want their policy purposes be served. They see artists not as eye-level partners in the search for an Anthropocenic future. This attitude tends to overlook the potential role of culture for political governance, for example, the culture of dialogue, the culture of conflict resolution, the management culture in associations and companies, the culture of recognition of achievements. This potential is, in our sense, a dormant seed.

Nevertheless, we see a whole series of recent grassroots initiatives in the field of art and culture. Cultural institutions in Germany are opening up to questions of sustainability, both in terms of programmatic openness and in terms of internal sustainable management. The German Cultural Council, the umbrella organisation of all cultural institutions, and the professional association Kulturpolitische Gesellschaft are taking up the sustainability agenda. They explicitly refer to the 2030 Agenda and the new climate movement. A group of filmmakers called changemaker.film is developing minimum standards for sustainability in the shooting and production of film images. Theatre makers, exhibitors and the music industry discuss how they can contribute to climate protection. Theatre plays deal with climate protection, readings and performances take up these topics.

Since 2001, the Sustainability Council very persistently provided impetus and impulses, whether through exhibitions and competitions, thematic concerts and dance actions. The council supports some 90 transformative projects on sustainability culture with funds allocated by Parliament for this purpose. Since 2001, the Council has pushed for the German government's sustainability strategy to include the special role of the arts, culture and creative industries (see Figure 11.2). The 2016 relaunch of the German Sustainability Strategy finally included this idea for the first time (Bundesregierung 2016, p. 45 f.). In its first-ever sustainability report, the federal government's Commissioner for Culture and the Media outlines activities related to the sustainability goals (BKM 2020).

A couple of exemplary short artworks powerfully illustrate the deep thinking that the arts can convey, contrary to the regular language, including science, used in the sustainability context. They succeed in replacing stale myths, connecting the individual with the social, and creating a sense of time in one's life and of the uniqueness of one's generation in transition.

→ involve more people than ever before

→ inform the debate

→ advance policies: energy transition, land use, resource efficiency, precaution

→ keeping the political momentum

→ make culture and arts a factor

→ institutionalise norms: German Sustainability Code

Figure 11.2 Main achievements of RNE

An irritating idea of time is the essence of a thought that stems from Gerhard Gundermann (1955–1998), the famous East German singer/songwriter, who used to hurry at night after giving concerts to his job as machinist excavating lignite in an open top mine. In his song titled "no time left" he once felt that his running for life makes him running against the course of life. He expressed the notion that in terms of environmental damage there is no time left, and for the survival of all one must change one's course of life. The India born young poetry slammer Aranya Johar, in her poem on sustainable development, is all about values and the ethics of collective impact. She addresses ordinary people as heroes who don't wait for crisis to save the day and the planet. Those heroes would make others heroes every day. An even more transformative operation is what Rainer Maria Rilke, the Austrian poet (1875–1926) had in mind when he described the transformative character of hope and expectations. He spoke about the future as a subject that when it enters us, transforms itself in us, long before it happens. In his letter number eight to a young poet, from August 12, 1904, he felt that when people think of what lies ahead in the future, this thinking changes human behaviour, and so, is shaping the future. Leonhard Cohen is known for the sentence: There is a crack in everything, that is how the light gets in. Cohen makes it clear: Optimism is a matter of how intensive one looks for the crack.

These examples stand for seeds of hope. They show creative, self-thinking and overcoming of obstacles. They show self-esteem that affects the course of life, and a collective notion of heroes. They emphasise the power of processes. Such lyrical proposals can only come from artists. They are, of course, very different from policies that seek to steer sustainable development essentially through goals and dystopian messages.

11.4 Seeds of iconic semiotics

Culture can transfer notions and myths into semiotic landmarks and make them evocative and attractive. Icons can also work in a negative sense, deepening exclusion and contributing to antagonistic enemy images. And they can evolve from one to the other. The debate on sustainability has seen a number of examples that lost their positive value. Now they are empty and obstructive shells, while new ideas germinate. An example is the semiotic image that picture sustainability as a triangle with three same-sized side length – ecology, economy, social – (like-wise as three pillars as a "house" of sustainability, or as three partly overlapping circles). In the early days of the political debate on sustainability it had some merits and stood for a shift in power. The GDP economy, it said, is no longer the exclusive yardstick for prosperity. Ecology and human rights are on a par with the economy. This radiated rationality, indicating that everything was under control: nothing happens or comes from outside. Sustainability rests in itself and is a final state that arises from an idea of redemption. Columns or circles of equal size pretend that power would be equally distributed. The point is: Nothing of this is true or even remotely likely to serve as a lighthouse vision. In reality, very much like in a Mikado-type game (those who moves first, will lose), the actors in Germany used the pillars, triangles and circles for absurd positional battles: Who is more important? Who would be named first?

Today, most organisations use the colourful tiles of the SDGs as a proxy for their commitment to sustainability. This is the right thing to do and is bearing fruit. In Germany, more people than ever before are familiar with the term sustainability. In 2001, that was only about 10 percent; today, it's over 80 percent. The message is clearer and more practical than ever before: it's about timeliness and processes and not about eternity and an end state; goals harbour conflict and that's why governance and gatekeeper, peace and the rule of law are important; goals must be adapted and that means legitimised nationally, a point for democracy. SDG tiles today stand for this positive belief, on one condition: They must help to overcome the delivery gap, the vicious circle of commitment and inadequate action. If the delivery gap is not closed in time, the tiles literally will fall off the wall.

Common But Differentiated Responsibilities (CBDR) are another one of these outdated, but still impactful iconic myths. This concept was enshrined in the United Nations Framework Convention on Climate Change and has had its merit as an international legal instrument – the first – to combat climate change. It recognised the wide disparity in economic prosperity and historical carbon load between countries. Today, the initial merit that has worn away. Among the countries of the world, there are not anymore only two distinct categories of national responsibility. All national greenhouse gas emissions are contributing to global warming. That makes responsibility indivisible and universal. Increasingly, the widest disparity in prosperity can be found within the countries, rather than between countries. For the high emitters, CBDR are another means of ducking behind the carbon poor. It is often misused to defend carbon claims and to stay with fossil industries.

The sustainability community is bad at developing such icons that have a coherent message and an appealing form. The often-used image of children's hands holding some soil and a young plant inside is nonsensical. It is naive. It does not represent reality and does not really stand for a vision. But it accurately reflects the movement's ability to create an image of the future and to present its own role critically and self-confidently. There is probably a very deep-seated reason for this. Sustainability policy is incapable of dealing with time, against the odds of a seemingly time-bound definition. In fact, there is no conceptual frame for future or transience or eternity. To cover this flaw, actions and symbols are shallowly tied to short-term campaigns or political goals, but not to the conditions of existence and a full understanding of the human-nature relationship. The power of aesthetics is not yet used. There is still much to do here.

11.5 Regrouping the players

Adapting to new issues always means gaining or losing significance, careers, money and power. Although this is a sometimes difficult institutional and individual challenge, regrouping is routine and happens over and over again. The task for sustainability is to understand its dynamic and to use it purposefully.

Sustainability is already changing the power structures within the business camp. The topic demands pre-market competition for concepts and minds. For companies this is an opportunity for differentiation and innovation. Social media are shifting the rules of the game in communication, also as regards sustainable management.

It is not long ago that corporates, citizen groups, marketing facilities felt not at all at ease with the term sustainability. For a long time, this was frowned upon and occasionally written in parenthesis, like to maintain a hygienic safety difference. Company bosses and advertisers considered the term a complete no-go. Environmental campaigners refrained from using this term in order to save their image as clean and green voice. For them all banks and manufacturers camouflage and deceive by greenwashing. Campaigners have repeatedly made this clear in Glasgow on the occasion of the climate COP with strong visual motifs.

Today, things are different. Rotten apple will always exist, but as a collective structure, greenwashing is over. Today's top managers believe that sustainability will be at least as important as digitisation or even more important in the next five years (Raffel and Wörner 2021). It can be assumed that large companies will need another 10 to 15 years to get down to full carbon neutrality. That would be a few years earlier than Germany as a whole. This is where the jobs of the future lie, especially if technological progress finds a corresponding social and cultural practice (Vassiliadis and Borgnäs 2020). Companies with good sustainability performance have, according to Accenture (2021), on average 3.7 times higher operating margins than companies that do not measure their performance or do not measure it thoroughly against relevant indicators.

Another driver of regrouping is foresight, usually (but not sufficiently) expressed by figures. One example is the order of magnitude of investments needed to neutralise Germany's carbon balance sheet. According to a study commissioned by the leading German bank KfW (Burret 2021) around 4.5 trillion euros will be needed until 2050. This corresponds to an annual average of 3.9 percent of the GDP. That sounds like a terrible sum and it leaves politicians and people stunned and it does not seem to motivate them. But then, as a comparison, the gross fixed capital formation amounts to almost 170 trillion euros, and the trend is upward. In 2016 alone, the manufacturing sectors invested 53.4 billion euros in research and development. Manufacturing contributed 23.4 percent to gross value added in Germany in 2017. In France the share is 12.7, in the UK 10.1 percent.

What this means: The investment costs into climate neutrality seem high, but they are not. To speak of peanuts would not be justified, but on the other hand, attributes such as high, costly, expensive are not justified either. Such high numbers need to be used to think about perspectives, and that sheds a different light on the issue and makes people feel like they're on the move again. Thinking has a seed component: Historically, no one even thought of calculating in advance the costs of the German *Wirtschaftswunder*, economic miracle, in order to be able to decide then whether one really wanted a Wirtschaftswunder. No one calculated what German reunification would cost in order to decide whether it would be better to leave the Wall standing. No one ever thought of making the introduction of cell phones dependent on what their infrastructure would cost. Only in the case of climate protection is there a tendency to over-exactly calculate everything. That is strange. For one thing, the usual financial ratios do not offer any systematic certainty anyway, because they ignore their social and environmental externalities. And why should it make sense to know the costs when it comes to climate protection of all things, with all its high physical uncertainties and global interdependencies? Why, of all things, should we leave a question of human survival to an economic calculation? And, even we do want to see calculations, why, of all things, are we calculating what we call investment costs, and not profits that will surely be made by profiteers, or societal prosperity?

The large number rhetoric is probably just an attempt at lobbying, a helpless one at that. It puts into figures only what is known. That is like the iconic picture of a person at night in the streets desperately searching for the lost bracelet. The bracelet he has lost in a dark passage, but the person searches under the light of the street lamps, because only there he can see anything at all. More relevant than what seems predictable and calculable is what it is not: indirect risks due to cluster effects in the economy, uncertainty due to misunderstood or failed regulation, burdens due to bureaucracy, and costs for health and preventive care, prosperity beyond numeric growth. Not to be neglected are the so-called transition risks – effects caused by the transformation itself – which can result from jumps in the carbon prices, green valuations of stocks and corporate bonds, or green race competition. This is especially striking when accompanied by ignorance on the part of the risk's owners.

Globally, big finance commits to re-wire the financial system taking climate into account. The finance industry is about to produce a unified sustainability reporting standard. Corporate reporting on sustainability is gaining momentum. Asset management companies are taking a close look at it. Companies are experimenting with fully expressing their corporate value including externalised social and environmental values (be they costs, losses or profits). Some are experimenting with navigating full value balancing systems (Quarta Vista 2020). Large corporations are pushing policymakers (in Germany) to adopt more ambitious and effective regulations for climate protection. This is a defining turn-around: While previously, the corporate thinking was aligned to avoiding regulation, today (not all, but) leading companies ask for more and better, legally binding regulation of markets. They go even beyond the carbon taxation and make it clear that circular economy and a climate-compliant business need active industrial policies, and they encourage government to leave the moral high ground of allegedly not interfering with market forces.

The European Union is committed to rearranging the financial industries and to support sustainable financing. Key is its so-called taxonomy that defines how green is green. Other players are the Network for Greening the Financial System of a group of central banks as well as the worldwide cooperation of audit institutions, the Supreme Audit Institutions. The European Corporate Sustainability Reporting Directive, CsrD, is expected to follow up on the CSR guideline that since 2014 requested big public companies to file sustainability reports. The mandatory reporting will also affect public companies in the EU. In Germany, this could be relevant for up to 3,000 of the total 18,500 public companies. The desired outcome of all this is a transparent, single-source report with approved data in synchronised availability and compliant with the Paris climate target relative to a 1.5 °C scenario. There is no doubt that the desired outcome is more than a necessary step, one of many. But market participants alone will not be enough, nor will the European Commission's role as facilitator. A different approach is needed, and this seed must come from the public. The costs or benefits a textile company calculates for using a ton of Mekong water cannot simply be left to the market. Even a complete market, whatever it may be, needs standard values and benchmarks for environmental consideration that the market itself can never provide. A mechanism for due diligence must specify basic environmental standards. This is the seed of change.

The US, the European Union and the UK, in an attempt to curve imperial infrastructure malpractice, propose an idea of building infrastructure that would simultaneously advance prosperity and combat the climate crisis, in line with the Sustainable Development Goals and the Paris Agreement.

Overall, leading companies and institutions have quite substantive answers to the SDGs; others are struggling to find them, a minority is actively denying the case. The answers may not yet be sufficient, but they provide a basis for public positioning. Companies that discover the strategic core of sustainability have an easier time with their purpose and are more attractive to young employees and more creative. Those companies that are not yet there are often stuck. Those who deliberately set themselves up against sustainability and want to continue their carbon empires will surely be punished either by the market or by regulation.

The redirection of value creation and financial flows to cutting carbon emissions is well underway, and the pledge of $100 billion annually in climate finance from countries is likely to be met in due course. Germany and Japan are the largest donors to this pledge. At least in Germany, the public is not really aware of the financial transfer. If this were the case, questions and debates will arise, especially with regard to an unreflective, superficial inflation of the term climate/sustainability investments. If everything is climate investment, nothing is. You have to look for the loopholes. One of the most important is corruption, a problem that unfortunately works against the good will of countries and companies but is not addressed because of a misconception of political correctness. The problem is not corruption as such, but that it is covered up by political correctness. Bhattacharya and Calland (2020), in an official report to UN Secretary General Antonio Guterres, state that while there is a great and morally motivated debate about the failure of the carbon-rich North because it has not yet raised the full $100 billion, no one is attempting to measure the factual impact of the portion of the funds of the annual $100 billion that has been spent so far. The most (climate) vulnerable regions of the world are also the most exposed to corruption and failing governance. The Corruption Perceptions Index compiled by Transparency International, which the German government has made part of the GSDS, impressively underscores this connection.

So it can be assumed that with private assets and taxpayers' money now increasingly flowing into investments in climate impacted countries, a large part of it will be lost in corrupt structures without anyone taking this into account. This is a bizarre point: while everyone is advocating science-based goals, no one seems to care about science-based efficiency of the funds used. The reason is that it is thought to be politically incorrect to raise the climate corruption issue, because it may look like an excuse to refrain from funding. This is the old and never-ending story of governance failure: silence prolongs and exacerbates the problem. Silence creates accomplices. Those who cling to this false belief and its political incorrectness fall deep into a creativity coma.

Only by speaking out about reality can the seeds for change sprout. Those who interpret sustainability as a process and in its diversity will reap from this seed. In Germany, this is the case for gatekeeping. Institutions are called upon to ensure clear content and verifiable criteria in the use of the term, and to separate serious from superficial proponents. It is obvious: In order to put the relationship between man and nature on the path of sustainability, the administrative and interest-led organisations will be the key. For these in particular, regrouping is necessary. But how?

ADDRESSING THE PRIVATE SECTOR

The Sustainability Code standardizes the voluntary reporting on non--financial performance that can be used by all organizations and enterprises regardless of their size or legal structure. It is stakeholder based and backed by a helpline, review and a databank by the Code Office. Since 2010 The Hub for Sustainable Finance – initiative of RNE triggered cooperation with the German Stock Exchang, and a major regrouping of German financial industry actors.

BOOSTING GRASS ROOTS

The German Sustainability Action Week brings together thousands of voluntary well crafted projects and initiatives, now part of the European Sustainable Development Week.
The Council boosts the culture of acknowledgement by inviting grass root project to take part in the RNE lineup of transformative projects and ideas. Since 2005.

MANAGING LOCAL COMMUNITIES

The four Regional Hubs for Sustainability Strategies (Regionale Netzstellen Nachhaltigkeitsstrategien – RENN) connect local activities and are boosting sustainability strategies. The Council runs a central steering capacity and organises the funding. Since 2016.

The RNE-invited Lord Mayor's strategic dialogue "Sustainable City", since 2010, produces interventions and strengthen the local levels voice in federal politics.

ADDRESSING CONSUMPTION AND PRODUCTION PATTERNS

The Sustainable Shopping Basket offers consumption alternatives and tips for sustainable living, e.g. food, clothes and sustainable money investment, since 2004, updated and enriched regularly. 2019 transferred to the regional RENN network.

INTERNATIONAL DIALOGUES AND NETWORKS

BRICS+G Dialogue on Sustainability and Growth with Brasil, Russia, India, China and South Africa, 2005.

SDGclub.Berlin, first conducted 2016 shortly after the passing of the 2030 Agenda. A second SDGclub.Berlin took place in 2019.

Peer Reviews with international high-level eminent peers, invited by the Federal Governmant to gauge the German sustainability approaches, and to ensure seamless continuation of top-level engagement, with the Council facilitating the review work and connecting it the German policy makers and communities. 2009, 2013, 2018.

European Networks of environmental and sustainability advisory councils (EEAC) plus support of the informal network of European public administrators and experts, European Sustainable Development Network (ESDN), both since 2002.

Global Forum for National SDG Advisory Bodies, initiated by RNE, Finland and UNDESA. Since 2019.

ON ROOTS AND CULTURE

Carl von Carlowitz Lectures: A series of eight "old style" academic lectures by renowned professors, launched in 2013 to commemorate the 300th anniversary of the publication of Sylviecultura oeconomica (1713) by Carl von Carlowitz. Transferred as a project to the Carl-von-Carlowitz Society, Chemnitz, and continued there.

The project Politics and Culture is a series of concerts, dance performances and art exhibitions held each year as part of the RNE's annual political summit conferences. The series featured performances by well-known artists and amateurs, school orchestras and professional musicians, all willing to explore sustainability issues and create unusual approaches.

ON HOW TO WALK THE TALK

The Council plans and implements conferences and events according to the criteria of sustainable event management. In case the criteria are not yet sufficient the Council uses his high prestigious summit conference for experiment, such as with a zero-food-waste practise, novel insect food, or tap water, buying-in of green power. Experiences are publicly reported, and shared with government officials.

Figure 11.3a–b Major RNE projects

In fact, some regrouping has already begun, exactly where there are cracks in the conventional organisation of government and advisories. Multistakeholder bodies are emerging that bring together diverse and even conflicting interests in a common framework for discussion and action. Over the past 20 years, they have developed sustainability criteria for coffee, tea, cocoa, palm oil, fish, vanilla, conflict minerals, and proteins, to name a few. Sophisticated analyses of these criteria now exist. Hemmati (2002) analyses how conflict becomes so entrenched that it is confirmed again and again. The conflict becomes the reason for the conflict. This is where a well-organised multistakeholder perspective can help. Guidelines and manuals describe how this perspective ideally works (Brouwer et al 2016; Dodds 2019; Strandenaes 2021).

Ideal concepts help with orientation, but idealisms are rarely implementation-ready. The political culture differs a lot between countries, policy cycles, and issues. Rather, people tend to experiment with different concepts. Germany's experience with the political commissioning of multistakeholder groups is good: the Ethics Commission "Secure Energy Supply" has so far had the greatest selective, transformative impact (nuclear phase-out). The German Council for Sustainable Development has the broadest and, in terms of permanence, greatest transformative impact (see Figure 11.3a–b). Other issues such as the coal phase-out or sustainability in agriculture also received valuable impetus by similarly composed commissions.

Measured against the past, all those transformations are astonishing. Measured against the material challenges of the Anthropocene, it is timid, too slow, and fragmentary. Leadership is still too sedately treading water.

We see a cumbersome, ill-favoured and often dysfunctional multilateral cooperation – while the pandemic, global warming and the 2030 Agenda prove the nation state to be overburdened. It is seen a remnant of the 19th and 20th centuries. And the belief that multilateralism can contain national egoisms has failed.

The major global political initiatives of recent years have certainly been shaped by the nation-state. It was the national level that gave form and substance to the Paris Climate Accords and the Sustainable Development Goals, as well as to the global vaccine alliance, GAVI. It is at the national level that parliaments and politicians act with direct responsibility and based on voters' decisions; it is here that civil society is strongest; it is here that cities and local communities act most strongly as catalysts and laboratories of progress. This is where things become concrete. This is why there is also a renaissance of nation-state competence and responsibility in globalisation.

The 2030 Agenda must be adapted and implemented nationally. National sustainability strategies, if set up correctly, give new impetus to multilateral thinking: it promotes cooperative action across national borders, along supply chains, according to ecosystem services and global common goods. It makes sense for the United Nation institutions to play a special role here.

Geopolitical tensions have so far not destroyed the 2030 Agenda. Contrary to some assumptions, UN member states continue to make ambitious commitments, and they are using the UN as a platform. These obligations are clearly insufficient, but they still serve an important function. They socialise the SDGs and enhance the HLPF as a multilateral platform. In 2019, UN members encouraged the secretary-general to organise an annual event to inspire action on the goals. This is what is now known as the SDG Moment during high-level UN General Assemblies. Leaders routinely describe the many crises and lament that they are off-track, that goals seem unattainable and the implementation gap appears to be widening. It's the biggest lament possible, and it sounds like a confession when there's no one in the room to give absolution. Acknowledging this delivery gap the UN secretary general repeatedly calls on all actors to bring more ambition and urgency to their implementation efforts in what he calls the decade of action and delivery. Javier Surasky (2021), the Argentine programme officer with the Global South think tank CEPEI, criticises that the capacity for data-driven decision-making is still limited and targets to build up capacity in developing countries are still unmet. For the Latin American countries, but generally true for the Global South, he observes that the statistical capacity receives less instead of more attention each passing year.

Knowing all this, it would be easy to lose hope. But to be hopeless is not an option, because we are not helpless. There are political options that may not be as disruptive, but certainly offer more than just small improvements. In our sense, they are seeds. First, redesigning the governance capacity of multilateral institutions would replace the workshops with a UN sustainable development institution empowered to establish binding soft rules and guidelines for conducting peer reviews, institutionalising multistakeholder processes, and interacting with the private sector, particularly in setting standards for calculating social and environmental losses and gains. Second, why shouldn't the International Court of Justice also serve as a supreme court of sustainability of global commons? Why shouldn't this court issue a kind of climate migration passport for people who must leave their homes to escape climate-related disasters? Third, an international effort to promote better-informed decision-making would be very useful. This would replace the current dysfunctional patchwork, and it would provide a science-based and data-driven background that people would not dare deny or put on the back burner. Fourth, a multilateral emergency response mechanism could provide organised and immediate assistance in the event of forest fires, pollution, floods, and other out-of-reach events that overwhelm national forces. This would fill the severe gap between financing humanitarian help and financing development cooperation and aid. And it could also strengthen the political concept of the responsibility to protect.

References and recommended reading

Accenture SE (2021) *Signals of Change: The Essential Radar That Leaders Need to See and Seize the Future, Series of Business Futures 2021*, www.accenture.com/_acnmedia/Thought-Leadership-Assets/PDF-5/Accenture-Signals-Of-Change-Business-Futures-2021-Report.pdf

Bhattacharya, Amar and Richard Calland (co-chairs of an independent expert group on climate finance) (2020) *Delivering the $100 Billion Climate Finance Commitment and Transforming Climate Finance*. Report to the United Nation Secretary General, www.un.org/sites/un2.un.org/files/100_billion_climate_finance_report.pdf

BKM (2020) *Nachhaltigkeitsbericht der Beauftragten der Bundesregierung für Kultur und Medien (BKM)*, www.bundesregierung.de/statisch/nachhaltigkeitsbericht-bkm/#2

Brouwer, Herman, Jim Woodhill, Minu Hemmati, Karèn Verhoosel and Simone van Vugt (2016) *The MSP Guide, How to Design and Facilitate Multi-Stakeholder Partnerships*. Wageningen: Wageningen University and Research, WCDI; Rugby: Practical Action Publishing, http://doi.org/10.3362/9781780446691; www.mspguide.org/sites/default/files/case/the_msp_guide_3rd_ed_2019_wcdi_brouwer_woodhill.pdf

Bundesregierung (2016) *German Sustainable Development Strategy, New Version 2016*, www.bundesregierung.de/resource/blob/998220/455740/7d1716e5d5576bec62c9d16ca908e80e/2017-06-20-langfassung-n-en-data.pdf?download=1

Burret, Heiko et al (2021) *Beitrag von Green Finance zum Erreichen von Klimaneutralität in Deutschland, Studie im Auftrag der Kreditanstalt für Wiederaufbau (KfW)*, www.kfw.de/PDF/Download-Center/Konzernthemen/Research/PDF-Dokumente-Studien-und-Materialien/Green-Finance-und-Klimaneutralitaet.pdf

Cepei (2021) *Comparative Analysis of the Voluntary National Reviews Presented by Latin America and the Caribbean Countries to the HLPF 2021*. Bogota: Cepei, https://cepei.org/en

Dodds, Felix (2019) *Stakeholder Democracy: Represented Democracy in a Time of Fear*, London: Routledge Publishers.

GlobeScan and SustainAbility Survey (2021) *Sustainability Leaders 2021*, https://globescan.com/wp-content/uploads/2021/07/GlobeScan-SustainAbility-Leaders-Survey-2021-Report.pdf

Hemmati, Minu (ed.) (2002) *Multi-Stakeholder Processes for Governance and Sustainability: Beyond Deadlock and Conflict*. London: Earthscan, www.wageningenportals.nl/sites/default/files/resource/multi_stakeholder_processes_for_governance_and_sustain ability_hemmati_2002.pdf

Homer-Dixon, Thomas (2020a) *Theory of Hope*, updated version for online posting, https://commandinghope.com/theory-of-hope/

Homer-Dixon, Thomas (2020b) *Commanding Hope: The Power We Have to Renew a World in Peril*. Knopf, Canada: Penguin Random House

Meadows, Donella, Jorgen Randers and Dennis Meadows (2004) *The Limits to Growth: The 30-Year Update*. White River Junction, VT: Chelsea Green Publishing Company.

Mulholland, E. (2018) *Cooperation Between Stakeholders and Policymakers in the Implementation of the SDGs: Overview of Activities and Practices in Europe*. ESDN Quarterly Report 50. Vienna: ESDN Office, October, www.esdn.eu/

Obama, Barack (2004) *Address to the 2004 Democratic National Convention*. Boston, www.americanrhetoric.com/speeches/convention2004/barackobama2004dnc.htm

Obama, Barack (2006) *The Audacity of Hope: Thoughts on Reclaiming the American Dream*. New York: Crown Three Rivers Press.

Porritt, Jonathon (2020) Hope in Hell. A Decade to Confront the Climate Emergency. London: Simon & Schuster UK.

Quarta Vista (2020) *Navigation System for Value Orientated Companies*, www.quartavista.de/en-gb

Raffel, Tobias and Stefan Wörner (2021) *From Attitude to Action: How Germany's CEOs Drive Sustainability*. A Study by FUTURIST, PIK und Bain & Company, www.bain.

com/de/insights/nachhaltigkeitsstudie-von-haltung-zu-handlung/ (English summary available)

Rand, Kevin L. and Jennifer S. Cheavens (2009) Hope Theory. In: Shane J. Lopez and C.R. Snyder (eds), *The Oxford Handbook of Positive Psychology* (2nd edn). Oxford, UK: Oxford University Press.

Snyder, C.R. (2002) Hope Theory: Rainbows in the Mind. *Psychological Inquiry* 13, no. 4, 249–275, https://www.jstor.org/stable/1448867

Strandenaes, Jan-Gustav (2021) *Stockholm+50 in 2022 – a Challenging Conference with a Forward-Looking Agenda.* Summary report of themes highlighted at Consultation 4 – of the Global Major Groups and Stakeholder's Forum, the GMGSF, Preparing for UNEA5.1, 2021, https://stakeholderforum.org/wp-content/uploads/2021/03/Stockholm50-Summary-Report-of-Themes-Highlighted-at-Stakeholder-Civil-Society-Consultation-4-for-UNEA-5.1–2021.pdf

Sunstein, Cass R. (2005) *Laws of Fear: Beyond the Precautionary Principle*. New York: Cambridge University Press.

Surasky, Javier (2021) *Comparative Analysis of the Voluntary National Reviews Presented by Latin America and the Caribbean Countries to the HLPF 2021*. Bogota: Cepei, https://cepei.org/en/documents/comparative-analysis-voluntary-national-reviews/

Time (1946) *Foreign Relations: Journey to Stuttgart.* The Speech on Hope by US Foreign Minister James Byrnes, Held in Stuttgart, September 16, http://content.time.com/time/subscriber/article/0,33009,888296,00.html

TSVCM, Taskforce on Scaling Voluntary Carbon Markets (2021), www.iif.com/tsvcm

United Nations (2019) *Political Declaration of the High-Level Political Forum on Sustainable Development*, HLPF, convened under the auspices of the General Assembly, September 24–25.

Vassiliadis, Michael und Kajsa Borgnäs (eds) (2020) Nachhaltige Industriepolitik. Strategien für Deutschland und Europa. Frankfurt and New York: Campus.

12 Unite behind politics

"Unite behind the science", presented by Greta Thunberg, summarises what carries most of the current young climate movement. The formula calls on policymakers to make scientific evidence the basis for decisions. It underscores the importance of choosing the best available science for effective climate action and the relevance of reasonable and sensitive science communication. With a focus on carbon metrics, loss and damages, adaptation, and nature-based solutions scientific research gets ever more important.

Unite-behind-Science was a timely wake-up call. However, it is still left a one-sided and half-hearted campaign which often makes the impulse only an internal experience of the usual convinced. Calling on policymakers would get stronger with a 180° feedback call that addresses the camp of climate activists and sustainability advocates. Today, it is equally important to get activists of all backgrounds to follow a Unite-behind-Politics. Politics is what counts.

If the German case of sustainable development policy shows one thing, it is that politics matters: Politics begin when you start telling a story about deep roots. Politics is about governing, about successes and a culture of failure, about the value you can place on democratic processes. Politics is when decisions are made about funding scientific research and innovation, long before an outcome is in sight. Politics is creating the moments to advance the cause of sustainability and it is about how you sow the seeds of change.

Governments and administrations will not come up with ambitious sustainability strategies unless they feel pressured and challenged by politically thinking and acting people on the street and in parliaments, in the neighbourhood and in enterprises who have a mandate or empower themselves. A notion of politics and policy is a must-have for strong activism. Cynicism in the sense of: politics always fails; all negotiations are a failure, greenwashing is everywhere, "they" do not understand, is the business of cowards. Instead of waiting wordily for the glorious silver bullet, it is a matter of courage to take the next, small steps. They may seem too little and even ugly, messy, and definitely unattractive. Small steps can go a long way, because taken together, they can make the big story. That is the case with Germany. When you look at Germany, you see quite a wide range of very different, but mostly small actions, on all level of public life. You see change in action. Transformation is normal. Transformation is moving forward in small

DOI: 10.4324/9781003242024-12

steps, each one of them may be insignificant, but in addition they show impact. Measured against the goals and urgent needs, the change is not enough. But we are no longer at ground zero. If this is being recognised more clearly, a couple of otherwise inconvenient truths become visible. This view, expressed in this book, is certainly in contrast to the prevailing political positions.

The make-believe worlds do not help.

The beliefs of our time either pretend that nothing is happening (except greenwashing), that Germany misses the necessary "moon shot" and that bitter lobbies are blocking away all opportunities, or – quite the opposite – that everything is happening too fast, that elite thinking is overpowering the regular mindset, that people are being left behind and that the economy is overstretched. Both are make-believe worlds, and to recognise this is a first truth that comes as unusual. Real transformation happens beyond this put-on negativism and its twin, the blind perfectionism.

The transformation does doesn't need such big words as moon shot or world saving and it certainly does not look like a heroic moon landing. And besides, some of the moon shooters are literally off the mark when you think of the aerospace company building a so-called gigafactory for electric cars in Germany. While the latter raves about building a completely self-sufficient life support system on Mars, on Earth they refrain from recirculating their factory's process water (groundwater being the most sensitive resource in the region). A good example of a take-away message from the German case: the bigger the words and the higher one rises to speak about morality from above, the further the protagonists are from the real change that is taking place as they speak to the stars.

Negotiation is possible, with people and with nature

There is a widespread belief that you basically cannot negotiate with nature. This belief makes us small, it even dwarfs us. Nature, the belief goes, speaks to us in terms of warming, depletion of the oceans, or decline in biodiversity, and we, humanity, have only the choice of either listening and obeying or withdrawing from nature's plan and thus perishing from Earth. That conviction dominates the mindset of many activists and politicians. They rightly point out that humanity is so stupid and conceited that it endangers its own existence. This is, obviously, the case and exactly why we need to transform our societies and let sustainability strategies guide the dynamics of change. But otherwise the proposition of non-negotiation is wrong. The human species is a part of nature and will remain so in times of the Anthropocene. So transformation is an internal dialogue within the whole of nature, and it is a dialogue about what humanity is and what it should be about. And further, the sentence misses the point that humanity is not a unity as long as some of its members hide behind the poor to disguise their exorbitantly large carbon footprint, including and especially the rich and powerful and the ecological jet-set. The answer to the non-negotiable proposition is democracy, and the quality of dialogue and multistakeholder processes.

Bold impulses are possible

And what about Germany's role in this world theatre? Square a military dwarf with an economic giant and take the history-loaded political culture into account and what you get is Germany's role in geopolitics and in the global climate policies in particular. Germany, with one percent of the world's population and 4.4 percent of the global wealth accounts for two percent of global CO2 emissions. That seems like a minor fraction. Nevertheless, the survival of climate protection will be decided in Germany.

Without the bold nuclear and coal phase-outs, motivated by a vision of sustainable development, there would hardly be any serious vision toward a green high-tech country anywhere. The German case encourages other country leaders to think bold.

Without Germany's pioneering role, the global boom in wind and solar energy would not exist. Smart regulation has sown these seeds in Germany, and the German electricity bill payer (not the taxpayer, it should be noted) has been footing the bill of more than 20 billion euros per year and continues to pay it with Europe's highest electricity prices (a circumstance that urgently needs reform). Then the market and innovators followed suit.

Without the German Chancellor talking the leaders of Japan, the US, the UK, Canada, Italy and France into supporting the notion of a global decarbonisation in their G7 meeting in Germany in June 2015, the tone for the Paris climate summit later that very year would have been different.

Without the long-developed and constantly supplemented governance of the DNA, other countries would not feel encouraged to work on their own strategies. And without them, the general acceptance of sustainability in Germany would not increase.

Germany certainly does not have the patent remedy for the Anthropocene, and no country in the world or institution does. German decisions are also not imitated everywhere. The nuclear phase-out is a case in point. But also the German regulation of an accelerated domestic climate protection has not been copied in the COP26 in Glasgow. The German focus on the technical solutions of renewable energies is sometimes seen as climate nationalism of an export nation, depending on the point of view, this is understandable.

Frankly, the German transformation process is not one of the most attractive and perfectly performed, considering how it meanders, stumbles and drags on. It sometimes looks nasty, nitty-gritty and almost a bit boring at times where you would expect it to look glorious, straightforward, and heroic. Unwise policy decisions combined with complacency among market participants have significantly delayed the energy transition but have not stopped it. The same applies to the circular economy and other sustainable production and consumption patterns. And that is exactly where the German Sustainability Strategy has its merits as a continuity script and a North Star mechanism, and as an impetus for the sustainability community. One needs very close insights into what is happening in policy circles, board rooms, and into jobs in manufacturing plants and in grass-roots initiatives and projects. Since 2001, the governance issue is under constant discussion and revision, mostly due to the pushing impact of gatekeeping institutions.

Just as a thought experiment, considering that Germany manages sustainability mainly through financial incentives and investment aid: What if one were to develop an alternative approach to sustainability that did not use financial incentives at all? Subsidies and other economic incentives would simply not be in the toolbox. After all, the transformation from horse-drawn carriage to car worked without incentives. This thought experiment shows what matters.

The big unspoken

The future will not be any kinder and more understanding than the present, why should it? No clairvoyants are needed to see a future that is less understanding. After all, the consequences of misguided developments in politics and society are increasing. More and more people experience as social conflict what in origin is an ecological issue. The times of benevolent apologies will quickly pass. The failure to achieve jointly set goals, combined with complacency, is becoming a burden on governments. In Germany, shopping is considered a form of leisure culture, regardless of income level. Buying beyond reason is the secret driver of the economy and the state's tax revenues. In general, Germans have a strong environmental awareness, if you think about waste separation, saving water at home or vegan/organic farmed food. But that's not the case at all when you look at the other side of the balance sheet: unprecedented amounts of packaging, fossil-fuel traveling, ever-larger cars, record obesity, food waste. Germans have a less sustainable lifestyle than they themselves often think. This may be true everywhere and for almost all halfway prosperous countries, at their respective consumption levels.

The logic of sustainable development requires a smart combination of technology, governance and culture, or in other words: policies and individual values, social common sense and individual change of habits. The combination counts. The only available instrument that at least begins to do justice to this combination is a sustainability strategy. It is at the beginning.

Index

Page numbers in *italics* indicate a figure on the corresponding page.

For Product Safety Concerns and Information please contact our EU
representative GPSR@taylorandfrancis.com
Taylor & Francis Verlag GmbH, Kaufingerstraße 24, 80331 München, Germany

www.ingramcontent.com/pod-product-compliance
Ingram Content Group UK Ltd.
Pitfield, Milton Keynes, MK11 3LW, UK
UKHW020140200926
475702UK00005B/831
* 9 7 8 1 0 3 2 1 4 9 5 2 3 *